Paths of Innovation

The first digital electronic computer, the ENIAC, was over 100 feet long, with 18,000 simultaneously functioning vacuum tubes. Now the compact PC is a ubiquitous feature of home and business. In 1903 the Wright brothers' airplane, held together with baling wire and glue, traveled a couple hundred yards. Today fleets of streamlined jets transport millions of people per day to cities worldwide. Between discovery and application, there is a world of innovation, of tinkering and improvements and adaptations. This is the world David Mowery and Nathan Rosenberg map out in their tour of the intersecting routes of technological change and economic growth in 20th-century America.

The authors focus on three areas of innovation that have dominated American technology in this century: the internal combustion engine, electricity, and chemistry. These three clusters of innovation are also highly research-intensive, allowing Mowery and Rosenberg to explore the importance of both "science" and technological innovation and improvement in realizing the economic consequences of technological advances. The cases of chemicals and the internal combustion engine also offer a lesson in the influence of geography and available resources on technological development.

Throughout their book, Mowery and Rosenberg demonstrate that the simultaneous emergence of new engineering and applied science disciplines in the universities in tandem with growth in the R&D industry and scientific research has been a primary factor in the rapid rate of technological change. Innovation and incentives to develop new, viable processes have led to the creation of new economic resources – which will, in turn, determine the future of American technological innovation and economic growth.

David C. Mowery is Professor at the Haas School of Business, University of California, Berkeley. He is coauthor with Nathan Rosenberg of *Technology and the Pursuit of Economic Growth* (Cambridge University Press, 1989), editor of *The International Software Computer Industry* (1996), and coeditor of the forthcoming *Sources of Industrial Leadership* (with Richard R. Nelson).

Nathan Rosenberg is Professor of Economics at Stanford University. He is the author of *Perspectives on Technology* (1976), *Inside the Black Box* (1983), and *Exploring the Black Box* (1994, all Cambridge University Press) and coauthor of *How the West Grew Rich* (with L.E. Birdzell, Jr.) and the title cited above with Professor Mowery, among other works.

Paths of Innovation

Technological Change
in 20th-Century America

DAVID C. MOWERY

NATHAN ROSENBERG

CAMBRIDGE
UNIVERSITY PRESS

PUBLISHED BY THE PRESS SYNDICATE OF THE UNIVERSITY OF CAMBRIDGE
The Pitt Building, Trumpington Street, Cambridge, United Kingdom

CAMBRIDGE UNIVERSITY PRESS
The Edinburgh Building, Cambridge CB2 2RU, UK   http://www.cup.cam.ac.uk
40 West 20th Street, New York, NY 10011-4211, USA   http://www.cup.org
10 Stamford Road, Oakleigh, Melbourne 3166, Australia

First published 1998

First paperback edition 1999

Printed in the United States of America

Typeset in Sabon 10.5/14 pt. in LaTeX 2_ε [TB]

A catalog record for this book is available from the British Library

Library of Congress Cataloging-in-Publication Data
Mowery, David C.
Paths of innovation : technological change in 20th century America
/ David C. Mowery, Nathan Rosenberg.
p.   cm.
ISBN 0-521-64119-5 (hardbound)
1. Technology–United States–History–20th century.   2. Internal
combustion engines–United States–History–20th century.
3. Electric engineering–United States–History–20th century.
4. Chemical engineering–United States–History–20th century.
I. Rosenberg, Nathan, 1927- . II. Title.
T21.M75   1998
609.73′09′04 – dc21   98-28901
CIP

ISBN 0 521 64119 5 hardback
ISBN 0 521 64653 7 paperback

For Moses Abramovitz and Richard Nelson,
esteemed colleagues and friends
who have taught us much.

Contents

Acknowledgments

The research underlying this monograph has absorbed us for much of the past five years. During that time we have incurred numerous intellectual debts. The greatest of these has been to Moses Abramovitz, a most valued colleague and mentor, whose incisive comments and constructive criticisms have greatly improved this monograph.

At various times during the protracted gestation of this volume, we have derived much benefit from discussions and correspondence with Ashish Arora, Timothy Bresnahan, Tom Cottrell, Paul David, Stanley Engerman, Alfonso Gambardella, Annetine Gelijns, Rose Marie Ham, Steven Klepper, Richard Nelson, William Parker, Scott Stern, and Walter Vincenti. We are especially grateful to Ralph Landau for guiding us through the intricacies of the chemical and petrochemical industries. He has been a splendid teacher. Victor Heffernan has availed us of his own vast experience in the application of the electric arc furnace to the steel industry and Peter Bridenbaugh has led us to a better understanding of the statistical data covering the aluminum industry.

We have also benefitted considerably from the research assistance provided by students at U.C. Berkeley and Stanford: Bonnie Chen, Thomas Cottrell, Amy Fisher, Rose Marie Ham, Guy Holburn, Christophe Lecuyer, Jeff Macher, Michael Preis, and Arvids Ziedonis.

We are pleased to acknowledge the support of the Alfred P. Sloan Foundation; the Andrew Mellon Foundation; the U.S.–Japan

Industry and Technology Program at the Haas School of Business, U.C. Berkeley; the California Policy Seminar; the Canadian Institute for Advanced Research; the Institute for Management, Innovation, and Organization, U.C. Berkeley; the Center for Economic Policy Research, Stanford University; and the Pine Tree Trust.

Since this is the second book we have written together,[1] we wish also to acknowledge the invaluable assistance of two technological innovations that were not available to us last time: email and the fax machine.

In the face of all of this invaluable assistance and support, we must insist upon our sole responsibility for errors and interpretive deficiencies in the pages that follow.

---

[1] See *Technology and the Pursuit of Economic Growth* (Cambridge University Press, 1989).

1

Introduction

A N EXAMINATION OF TECHNOLOGICAL INNOVATION in the 20th-century U.S. economy must naturally begin in the 19th century. An appropriate starting point is Alfred North Whitehead's observation, in *Science and the Modern World*, that "The greatest invention of the 19th century was the invention of the method of invention" (p. 98). Whitehead understood that this invention involved the linking of new scientific knowledge to the world of artifacts. But he also understood that this linkage was not easily achieved, because a huge gap typically exists between a scientific breakthrough and a new product or process. The sentence just quoted is well known, but equally important is the less famous observation that immediately followed it:

> It is a great mistake to think that the bare scientific idea is the required invention, so that it has only to be picked up and used. An intense period of imaginative design lies between. One element in the new method is just the discovery of how to set about bridging the gap between the scientific ideas, and the ultimate product. It is a process of disciplined attack upon one difficulty after another.

Whitehead's statement serves as a valuable prolegomenon in at least two respects to this volume's discussion of technology in the 20th century. First, a distinctive feature of the 20th century was that the inventive process became powerfully institutionalized and far more systematic than it had been in the 19th century.[1] This institutionalization of inventive activity meant that innovation proceeded in increasingly close proximity to organized research in the 20th century. Of course, this research was not confined, as Whitehead appreciated, to the realm of science, much less to scientific research of a fundamental nature. But Whitehead's observation is apposite in another respect as well. For all its reorganization and institutionalization, the realization of the economic impacts of 20th-century scientific and technological advances have required significant improvement and refinement of the products in which they are embodied. This process of incremental learning, modification, and refinement, along with the often prolonged process of adoption of these new technologies, means that even in this technologically revolutionary century, realization of the economic effects of new technologies requires considerable time.

This point needs to be emphasized because there is not always time or space to develop it in detail within the scope of a short volume. Inventions, when they are first introduced or patented, are typically very far from the form that they embody when they eventually achieve widespread diffusion; or, to put it differently, it is the improvements that they undergo that finally lead to widespread diffusion. The Wright brothers' achievement of heavier-than-air flight at Kitty Hawk in 1903 was a great technological accomplishment, even though the clumsy contraption was held together with struts, baling wire, and glue, and the total distance traveled was just a couple hundred yards. It required thousands of improvements, small and large, over fully a third of a century, before regularly scheduled intercity flights became common with the introduction of the DC-3 in 1936. The first digital electronic

---

[1] Schumpeter (1942) noted that "innovation itself is being reduced to routine. Technological progress is increasingly becoming the business of teams of trained specialists who turn out what is required and make it work in predictable ways" (p. 132).

computer, the ENIAC, was over 100 feet long and required the simultaneous functioning of no less than 18,000 vacuum tubes when it was introduced in 1945. Today an instrument with vastly superior capabilities can easily be held in one's hand or even carried in a pocket.

Moreover, many intermediate steps must be completed before the commercialization of such innovations. In many cases, ancillary inventions or improvements, frequently from other industries, are needed; new products must be redesigned for greater convenience, and cost-reducing changes are necessary to render them more affordable; further adaptations are necessary as consumers discover new unanticipated uses; production facilities need to be reorganized to adapt to the idiosyncratic production requirements of the new product. The time required for all these complementary developments to emerge is measured typically in years and not infrequently in decades.

Although considerable time is typically required for the economic effects of technological innovation to be felt, these effects are profound. Not until the 20th century had run more than half of its course did economists develop a fuller appreciation of the extent to which economic growth was a consequence of the process of technological change. One important reason for this delay was a simple one. An *economic* approach to economic growth is feasible only after certain magnitudes have been reduced to quantitative expression. The measurement of economic growth itself requires a conceptual framework and a methodology for aggregating the myriad of activities that make up the daily life of a large economy. The development of national income and national product accounts were products of the 1930s and 1940s, when a number of scholars developed estimates of long-term changes in the inputs to economic activity as well as the estimates of output. The availability of such data set the stage for the "discovery" in the 1950s of the contribution of technological change to long-term economic growth in the United States.

Once reasonably reliable estimates of the growth of inputs and outputs were available, it seemed natural to ask how much of the long-term growth in output could be accounted for by the estimated growth in inputs. The two most influential studies, by Abramovitz (1956) and Solow (1957), employed different methods, examined different time periods, and measured the economy's output in different ways. But these studies agreed on a very important conclusion: No more than 15% of the measured growth in U.S. output in the late 19th century and the first half of the 20th century could be accounted for by the growth in measured inputs of capital and labor. The strikingly large "residual" of 85% suggested that 20th-century American economic growth was overwhelmingly a matter of extracting more output from each unit of input into economic activity, rather than merely utilizing more inputs.

A prime candidate in accounting for the residual was technological change, although the residual necessarily captured all the growth in output that could not be attributed to growth in measured inputs of capital and labor. Incautious analysts drew the conclusion that the growth in the stock of capital did not make an important contribution to economic growth. But the contributions of technological change to economic growth rarely are independent of investment, because most new technologies need to be embodied in the capital goods that are the vehicles for their introduction. Most new technologies enter the stream of economic life only as the result of an investment decision. At the same time, however, the character of technological change, especially any factor-saving or -using bias associated with it, influences the demand for physical and human capital investment.

There is another connection between technological change and 20th-century U.S. economic growth. Kuznets pointed out in 1930 that technological innovation frequently creates entirely new industries devoted to the production of new products (see also Burns 1934). These new industries typically grow rapidly in their early stages and then experience retardation in their growth rates as their markets reach saturation. Because older final goods are likely to face lower demand elasticities, further cost-reducing process innovations have relatively small effects on demand for the product. The rate of growth of the entire economy is the summation of the growth rates of its component industries, which means that a high rate of aggregate growth requires that declining rates of growth in mature industries be offset by the more rapid growth rates of new industries associated with new technologies. In other words, sustained economic growth reflects a continuous shift in the economy's product and industry mix.[2]

Although insightful, Kuznets's statement tends to understate the importance of the adoption of new technologies by mature industries, which has sparked productivity growth and even the appearance of new products (e.g., synthetic-fiber radial tires) in these industries. In fact, many older industries have experienced significant productivity growth as a result of the *intersectoral flow of new technologies*. This intersectoral flow is a fundamental characteristic of 20th-century innovation in the U.S. economy – for example, innovations in the chemicals and electronics industries have been truly pervasive, being incorporated into a staggering array of consumer and capital goods. In addition, the rise of the automobile and commercial aircraft industries significantly increased the demand for advanced products (e.g., jet fuel, composite materials, gasoline) from other industries, thereby creating additional incentives for increases in scale and efficiency. The importance of this intersectoral flow of technologies is one reason

---

[2] "[A] sustained high rate of growth depends upon a continuous emergence of new inventions and innovations, providing the bases for new industries whose high rates of growth compensate for the inevitable slowing down in the rate of invention and innovation, and upon the economic effects of both, which retard the rates of growth of the older industries. A high rate of over-all growth in an economy is thus necessarily accompanied by considerable shifting in relative importance among industries, as the old decline and the new increase in relative weight in the nation's output" (Kuznets 1959, p. 33).

that we focus the discussion in this book on a few broad classes of technologies that have influenced innovation and growth throughout the U.S. economy.

The international flow of technology has also been important to U.S. economic growth. No treatment of technological change in the 20th-century United States can overlook the importance of this economy's technology imports and exports. Although the United States was overwhelmingly an importer of foreign technology during its early history (Rosenberg 1972), by 1900 it had become a considerable exporter of industrial and agricultural technologies. In fact, the United States had begun to export specialized machine tools as early as the 1850s. A collection of such tools was shipped to the Enfield Arsenal in Great Britain, where they laid the basis for the large-scale manufacture of firearms made of interchangeable parts – a technology that the British referred to as "the American system of manufactures."

At the same time, however, the United States imported a range of industrial technologies that, as of 1900, had not been mastered in this country. For example, German industry dominated the manufacture of organic synthetic dyes in the late 19th century. Indeed, the technological position of the U.S. economy before the Second World War bore more than a superficial resemblance to the situation of Japan in the 1960s and 1970s. During this period, U.S. firms had few equals in their ability to exploit (and often, improve) technologies that had been invented abroad. But until the Second World War, America's role in the world of science was in no way on a par with the leadership position that it had established earlier in the century in numerous realms of industrial and agricultural technology.

The general point is that, even at the beginning of the 20th century, the world economy was already well on its way toward globalization. Indeed, by some measures (such as the share of gross national product [GNP] accounted for by exports) the U.S. economy at the turn of the century (1900–10) was more open to international influences than at any time during the 1910–70 period (see Davis et al. 1972). By 1900, United States–based multinational firms, such as Singer Sewing Machine and International Harvester, were already playing a significant role in the international transfer of new technologies. The great improvements in the realms of transportation and communication were, of course, the main technological facilitators in the increasing flow of technological knowledge, as well as goods and capital, across political borders. Two disastrous world wars, the most severe depression in history, and the protectionist policies that separated those wars served to interrupt the trend toward globalization for several decades.

Since World War II, new international institutions, such as Bretton Woods and General Agreement on Tariffs and Trade (GATT), and, more recently, the World Trade Organization (WTO), have reduced barriers to the exchange of goods and technological knowledge. Indeed, the last third of the 20th century has witnessed the emergence of an increasingly dense network of interfirm relationships – international joint ventures and strategic alliances of all sorts – that contribute to more rapid international flows of technology (see Mowery 1988). The spectacular improvements in the information technologies that unite this international network have lent to the term "globalization" a vastly expanded significance over its meaning at the beginning of the century.

The discussion in the preceding paragraphs has been rather abstract, with its references to technological change in the context of "inputs," "outputs," and growth. In fact, technological change in the present century has produced a wide range of new products and manufacturing processes. The last decade of the 20th century differs from the last decade of the 19th, not primarily in terms of larger quantities of goods that already existed in the 1890s; rather, the distinctive differences consist in the present availability of goods for which there was no close equivalent a century ago.

An attempt to describe the new technologies of the 20th century would quickly take on the dimensions of a full shelf of

Sears Roebuck catalogues and trade journals – and even then these two abundant sources of information would fail to disclose much of what was distinctive about 20th-century technology. We focus instead on three clusters of innovation that have dominated American technology in the 20th century, using their development as the basis for a more general treatment of the central features of 20th-century U.S. innovation and its economic impacts. The three clusters are the internal combustion engine, chemistry, and electricity and electronics (we consider electricity and electronics as a single cluster).

These clusters have a number of common characteristics. They are pervasive – in fact, their economic impact has been more pervasive than is generally realized. Moreover, they are highly research-intensive, particularly if the term "research" is interpreted broadly and not confined to fundamental research at the frontiers of science. This is really no more than an acknowledgment that we employ the term to include all the components of what is now commonly referred to as *R&D* (*research and development*). Major new technologies have by no means always been dependent on new scientific knowledge. Innovation has throughout this century drawn on existing technological knowledge as much as it has on "science," and in some celebrated cases, technological innovations have appeared in advance of the scientific theories that explain their performance or design. Moreover, realization of the economic consequences of these technological advances has typically required considerable refinement and improvement of the crude early versions of the products that incorporate them.

The development within the United States of at least two of these three technology clusters, chemicals and the internal combustion engine, has also been influenced by this nation's unique geographic structure and resource endowment. The vast distances that goods and travelers must cover within the United States gave an impetus to the development and adoption of technologies that could shorten travel times, reduce transportation costs, and increase reliability of communication – in the 19th century, these were the telegraph and railroad (plus the older technology of canals), and in the 20th they were the automobile and the airplane.

The U.S. resource endowment, with its abundant supplies of raw materials – in particular, petroleum – also meant that the development of the internal combustion engine and the U.S. chemicals industry followed a resource-intensive trajectory. This characteristic of U.S. technological innovation reflects a more general phenomenon, the path-dependent nature of the innovation process. The initial conditions under which an innovation appears and is refined for economic exploitation exert a powerful influence over the types of knowledge required for its exploitation, the types of knowledge generated from its exploitation, and the evolutionary path followed by the technology.

Another distinctive feature of the history of innovation in the 20th-century American economy is the institutionalization of the innovation process that occurred during this period. In the late 19th century, industrial enterprises began to organize systematic programs of in-house R&D. The emergence of these industrial research laboratories in the U.S. economy occurred in parallel with the growth of new engineering and applied science disciplines in the universities. Indeed, all of our technological clusters are characterized by a shifting "division of labor" among private industry, universities, and government in R&D performance and funding. The structure of the U.S. R&D system that has spawned these waves of innovations underwent dramatic change during the first eight decades of this century, and since 1989, the end of the Cold War and economic globalization have triggered yet another wave of restructuring.

Before turning to our detailed discussion of 20th-century innovation, a general observation is in order. We have emphasized, so far, the role played by technological change in generating long-term growth in productivity and, therefore, in incomes. But the relationship between income and technological change is obviously a two-way street, even though the following pages travel on only one side of that street.

As a society's income rises, the composition of demand changes, and along with those changes in demand the profitability of inventions in different sectors of the economy changes as well. The vast commitment of resources in the United States today to the development of technologies aimed at recreational and leisure-time activities would have been inconceivable, or at least intolerable, during Thomas Jefferson's presidency, when the American workforce was overwhelmingly employed in food production. Similarly, if one is to account for American leadership, early in the 20th century, in the exploitation of the automobile, it is not sufficient to invoke the "wide open spaces" of a huge continental nation; the income levels that created a huge potential market for a "personalized" form of transportation were also essential.

Schmookler (1962, pp. 226–227) expressed this point well:

> It seems almost obvious, to this writer at least, that the automobile came when it did more because of economic and social changes than because of technological change as such. In the first place, in the automobile, prestige, flexibility, privacy, recreation, and utility are combined in ways which only an individualistic high-per-capita-income society could afford or develop. (The so-called bicycle craze of the 1890s was part of the same phenomenon.) The automobile, after all, did not and has not revolutionized life in low-income India or China. Its effects have been confined primarily to the United States and other industrialized countries, roughly in proportion to income. A good case can be made for the contention that among the indispensable conditions for the coming of the automobile age were relatively high levels of income, at least for the middle-income classes, and an individualistic society.

2

The Institutionalization of Innovation, 1900–90[3]

AS WE NOTED IN THE INTRODUCTORY CHAPTER, no account of technological innovation in the 20th-century U.S. economy can confine itself to a discussion of specific sectors or technologies. Another central element in the evolution of all industrial economies during this century was the transformation of the structure and organization of the innovation process. Like many other important technological advances in these economies, the development of organized industrial research was pioneered in Western Europe during the 1870s by German chemicals firms. U.S. industrial firms in chemicals and other industries quickly emulated this development, however, and by the 1920s, U.S. firms were, collectively, the leading industrial employers of scientists and engineers.

The U.S. R&D system that originated in the early 20th century has undergone profound structural change during this century. This structural change has two broad components. The first is the rapid exploitation by U.S. firms of the "invention of the art of invention" pioneered in Germany. A second, related feature of the evolution of the U.S. R&D system during this century is the shifting roles of industry, government, and universities as funders and performers of R&D. The magnitude of the shifts in importance among these three sectors within the 20th-century United States may well

---

[3] Portions of this chapter draw on Mowery (1995).

exceed that associated with any other industrial economy. The post-war R&D system, with its large, well-funded research universities and federal research contracts with industry, had little or no precedent in the pre-1940 era and contrasted with the structure of the R&D systems of other postwar industrial economies. In a very real sense, the United States developed a postwar R&D system that was internationally unique. On the other hand, the changes since 1989 in the international political environment that influenced so much of the postwar growth of the U.S. R&D system will continue to have profound consequences for the system's structure and international uniqueness.

The Origins of U.S. Industrial Research

The growth of U.S. industrial research was an important part of the restructuring of U.S. manufacturing firms during the late 19th and early 20th centuries. The in-house industrial research laboratory first appeared in the German chemicals industry during the 1870s (Beer 1959), and a number of U.S. firms in the chemicals and electrical equipment industries had established similar facilities by the turn of the century.

The growth of industrial R&D in both the United States and Germany was influenced by advances in physics and chemistry during the last third of the 19th century, which created considerable potential for the profitable application of scientific and technical knowledge. The original investments in industrial R&D were made by German firms seeking to commercialize innovations based on the rapidly developing field of organic chemistry. Many of the earliest U.S. corporate investors in industrial R&D, such as General Electric and Alcoa, were founded on product or process innovations that drew on recent advances in physics and chemistry. But change in the scientific and technological knowledge base alone cannot explain the growth of U.S. industrial R&D. New technical opportunities influenced the decision to invest in industrial R&D, but they cannot explain the growth in the share of R&D activity that occurred within the boundaries of the firm.[4]

The corporate R&D laboratory brought more of the process of developing and improving industrial technology into the boundaries of U.S. manufacturing firms, reducing the importance of the independent inventor as a source of patents (Schmookler 1957). In industries such as steel and meatpacking, materials inspection and testing facilities, many of which were established as the scale of production plants grew in the late 19th century, gradually expanded their responsibility for process and product innovation (Mowery 1981; Rosenberg 1985). But the in-house research facilities of large U.S. firms were not concerned exclusively with the creation of new technology. They also monitored technological developments outside of the firm and advised corporate managers on the acquisition of externally developed technologies.

U.S. Antitrust Policy and the Origins of Industrial Research

The structural change in many large U.S. manufacturing firms that underpinned investment in industrial research was influenced by U.S. antitrust policy. By the late 19th century, judicial interpretations of the Sherman Antitrust Act had made agreements among firms, for the control of prices and output, targets of civil prosecution. The merger wave from 1895 through 1904, particularly the surge in mergers after 1898, was one response to this new legal

---

[4] A substantial network of independent R&D laboratories provided research services on a contractual basis throughout the formative years of industrial R&D in the United States. These contract research organizations' share of total R&D employment, however, declined during the first half of the century. Moreover, many of their clients were firms with in-house R&D facilities, suggesting that contract and in-house R&D were complements rather than substitutes (Mowery 1983).

environment. Because informal and formal price-fixing and market-sharing agreements had been declared illegal in a growing number of cases, firms resorted to horizontal mergers to control prices and markets.[5]

The Sherman Act's encouragement of horizontal mergers ended with the Supreme Court's 1904 *Northern Securities* decision, but the influence of antitrust policy on the growth of industrial research extended beyond its effects on corporate mergers and remained important long after 1904. The U.S. Justice Department's opposition to horizontal mergers that lay behind *Northern Securities* caused large U.S. firms to seek alternative means for corporate growth. The threat of antitrust action that resulted from their dominance of a single industry led these firms to diversify into other areas. In-house R&D contributed to diversification by supporting the commercialization of new technologies that were developed internally or purchased from external sources. Threatened with antitrust suits from state as well as federal agencies, George Eastman saw industrial research as a means of supporting the diversification and growth of Eastman Kodak (Sturchio 1988, p. 8).[6] The Du Pont Company

---

[5] See Stigler (1968). The Supreme Court ruled in the *Trans Missouri Association* case in 1898 and the *Addyston Pipe* case in 1899 that the Sherman Act outlawed all agreements among firms on prices or market sharing. Data in Thorelli (1954) and Lamoreaux (1985) indicate an increase in merger activity between the 1895–98 and 1899–1902 periods. Lamoreaux (1985) argues that other factors, including the increasing capital intensity of production technologies and the resulting rise in fixed costs, were more important influences on the U.S. merger wave, but her account (p. 109) also acknowledges the importance of the Sherman Act in the peak of the merger wave. Lamoreaux also emphasizes the incentives created by tighter Sherman Act enforcement after 1904 for firms to pursue alternatives to merger or cartelization as strategies for attaining or preserving market power.

[6] Sturchio (1988) discussed the foundation of Eastman Kodak's R&D laboratory in the following terms, noting the important influence of state, as well as federal, antitrust policy: "Most important, however, was probably Eastman's increasing concern over the rise in antitrust sentiment on both the state and national

used industrial research to diversify out of the black and smoke-less powder businesses even before the 1913 antitrust decision that forced the divestiture of a portion of the firm's black powder and dynamite businesses (Hounshell and Smith 1988, p. 57).[7]

Although it discouraged horizontal mergers among large firms in the same lines of business, U.S. antitrust policy through much of the pre-1940 period did not discourage efforts by these firms to acquire new technologies from external sources. Many of Du Pont's major product and process innovations during this period, for example, were obtained from outside sources, and Du Pont further developed and commercialized them within the U.S. market (Mueller 1962; Hounshell and Smith 1988; Hounshell 1996).[8]

levels. As early as 1903 Eastman Kodak had faced state action against its alleged monopoly position in the photographic supply industry, and by 1911 there were signs that Kodak might also fall under Justice Department scrutiny as Standard Oil and American Tobacco recently had. ... Eastman, like his counterparts at other large corporations, saw in research the solution to new restraints to traditional competition created by antitrust measures. If mergers and horizontal combinations would no longer be allowed, research and development could lead to continued growth through the discovery of new markets and new businesses in a manner consonant with the Progressive rejection of traditional 'big business' practices" (p. 8).

[7] The Du Pont Company's research activities began to focus on diversification out of the black and smokeless powder businesses even before the antitrust decision of 1913 that forced the divestiture of a portion of the firm's black powder and dynamite businesses. Discussing Du Pont's early industrial research, Hounshell and Smith (1988) argue that "Du Pont's initial diversification strategy was based on utilizing the company's plants, know-how, and R&D capabilities in smoke-less powder (i.e., nitrocellulose) technology. The goal was to find uses for Du Pont's smokeless powder plants because political developments in Washington after 1907 [Congressional restrictions on procurement by the Navy of powder from "trusts"] signaled a significant decline, if not end, to Du Pont's government business" (p. 57).

[8] The research facilities of AT&T were instrumental in the procurement of the "triode" from independent inventor Lee de Forest, and AT&T researchers advised senior corporate management on their decision to obtain loading-coil technology from Pupin (Reich 1985). General Electric's research operations monitored

Writing in the early 1940s, Schumpeter argued in *Capitalism, Socialism and Democracy* that in-house industrial research had supplanted the inventor-entrepreneur (a hypothesis supported by Schmookler [1957]) and would reinforce, rather than erode, the position of dominant firms. The data on research employment and firm turnover among the 200 largest firms suggest that from 1921 through 1946, at least, the effects of industrial research were consistent with his predictions.[9] Mergers, management reorganization, and the development of giant industrial firms in the U.S. economy during the 1890–1920 period were associated with increased stability in market structure within manufacturing and a decline in firm turnover (Edwards 1975; Kaplan 1964; Collins and Preston 1961). Higher levels of R&D employment were associated with lower probabilities of displacement of firms from the ranks of the largest 200 U.S. manufacturing firms during the 1921–46 period (Mowery 1983). To the extent that federal antitrust policy motivated industrial research investment by large U.S. firms before and during the interwar period, the policy paradoxically may have aided the survival of these firms and the growth of a relatively stable, oligopolistic market structure in many U.S. manufacturing industries.

foreign technological advances in lamp filaments and the inventive activities of outside firms or individuals and pursued patent rights to innovations developed all over the world (Reich 1985, p. 61). The Standard Oil Company of New Jersey established its Development Department precisely to carry out development of technologies obtained from other sources, rather than for original research (Gibb and Knowlton 1956, p. 525). Alcoa's R&D operations also closely monitored and frequently purchased process innovations from external sources (Graham and Pruitt 1990, pp. 145–147).

[9] Interestingly, and in contrast to the usual formulation of one of the Schumpeterian "hypotheses," these results suggest that firm conduct (R&D employment) was an important influence on market structure (turnover). They are also broadly consistent with the results of studies of more recent data on the market structure–R&D investment relationship that suggest that structure and R&D investment are jointly determined (Levin, Cohen, and Mowery 1985).

The Role of Patents in the Origins of U.S. Industrial Research

The effects of U.S. antitrust policy on the growth of industrial research were reinforced by other judicial and legislative actions in the late 19th and early 20th centuries that strengthened intellectual property rights. The Congressional revision of patent laws in 1898 extended the duration of protection provided by U.S. patents covering inventions patented in other countries (Bright 1949, p. 91).[10] The Supreme Court's 1908 decision (*Continental Paper Bag Company v. Eastern Paper Bag Company*) that patents covering goods not in production were valid (Neal and Goyder 1980, p. 324) expanded the utility of large patent portfolios for defensive purposes.

Other Congressional actions in the first two decades of this century increased the number of Patent Office examiners, streamlined internal review procedures, and transferred the Office from the

---

[10] The original provision of the U.S. patent law and the timing of the changes in it are indicators of the growing role of the United States as a source of industrial technology, a change from its historic status as a "borrower" of industrial technology from foreign sources. As Bright (1949, p. 88) notes, "The U.S. patent laws at that time [the 1890s] contained a provision that an American patent was valid only as long as the shortest-lived patent in a foreign country, if the foreign patent had been issued first. The Canadian patent [for Edison's carbon filament] was declared invalid by the Canadian Deputy Commissioner of Patents, on February 26, 1889, for non-compliance with Canadian statutes regarding manufacture and importation. If that decision had been allowed to stand, the American patent would probably have become void also, since the Canadian patent had been granted before the American patent." Pressure from U.S. firms that had become patentholders led to revision in this statute: "The American patent laws were revised as of January 1, 1898 to include a provision that domestic applications for patents could be filed anytime within seven months of the earliest foreign application without prejudicing the full seventeen-year term of the American patent, regardless of its date of issue. The revision resulted in large part from agitation created during the early nineties when the Edison patent and a few fundamental patents in other industries were cut short before their full terms. The modification had an important bearing on the length of patent protection in incandescent lighting after that date" (Bright 1949, p. 91).

Interior to the Commerce Department, an agency charged with representing the interests of U.S. industry (Noble 1977, pp. 107–108). These changes in Patent Office policy and organization were undertaken in part to improve the speed and consistency of procedures through which intellectual property rights were established. Stronger and clearer intellectual property rights facilitated the development of a market for the acquisition and sale of industrial technologies. Judicial tolerance for restrictive patent licensing policies further increased the value of patents in corporate research strategies.

Although the search for new patents provided one incentive to pursue industrial research, the impending expiration of these patents created another important impetus for the establishment of industrial research laboratories. Both American Telephone and Telegraph and General Electric, for example, established or expanded their in-house laboratories in response to the intensified competitive pressure that resulted from the expiration of key patents (Reich 1985; Millard 1990, p. 156). Intensive efforts to improve and protect corporate technological assets were combined with increased acquisition of patents in related technologies from other firms and independent inventors.

Patents also enabled some firms to retain market power without running afoul of antitrust law. The 1911 consent decree settling the federal government's antitrust suit against General Electric left their patent licensing scheme largely untouched, allowing the firm considerable latitude in setting the terms and conditions of sales of lamps produced by its licensees, and maintaining an effective cartel within the U.S. electric lamp market (Bright 1949, p. 158).[11] Patent licensing provided a basis for the participation by General

---

[11] Bright (1949, p. 158) noted that "What the [1911 consent] decree did not require was of equal importance. No restriction was placed upon a manufacturer's right to acquire patents to fortify his interests. Moreover, the decree expressly stated that patent licenses might specify any prices, terms, and conditions of

Electric and Du Pont in the international cartels of the interwar chemical and electrical equipment industries. U.S. participants in these international market-sharing agreements took pains to arrange their international agreements as patent licensing schemes, arguing that exclusive license arrangements and restrictions on the commercial exploitation of patents would not run afoul of U.S. antitrust laws.[12]

Changes in the structure of the U.S. intellectual property system in the early 20th century, as well as the treatment of intellectual property by the judiciary, thus enhanced firms' incentives to both internalize industrial research and to invest in the acquisition of technologies from external sources. Against the backdrop of tougher federal enforcement of antitrust statutes, judicial decisions affirming the use of patents to create or maintain positions of market power also created additional incentives to pursue in-house R&D. Stronger, more consistent intellectual property rights also improved the operation of a market for intellectual property,

---

sale desired, although they could not fix resale prices. That permission left an enormous opening for continued control over the incandescent-lamp industry by General Electric, and the industry leader took full advantage of it in later years. Because the GEM [General Electric Metalized], tantalum, and tungsten lamps were rapidly replacing the ordinary carbon lamp, an open market for carbon lamps was not of much importance. General Electric's control over prices charged by its licensees was not seriously affected, and it retained its patent monopoly over the new types of lamps."

[12] Discussing the ICI-Du Pont 1929 Patents and Processes agreement, Taylor and Sudnik (1984, p. 126) argue that "Although both parties hoped to establish an understanding within which their home markets would be protected and provisions would be made for an orderly exploitation of new chemical technologies, Du Pont took pains to make the agreement conform to American antitrust laws as they were understood in 1929. John K. Jenney, secretary of the Du Pont foreign relations committee at the time, maintained that: 'It was the opinion of our lawyers that it was perfectly legal to relate commercial restrictions to patents... It was legal to license a patent or a secret process on an exclusive basis, which had the effect of preventing the export by the grantor of the patent license of a product covered by that patent or secret process.'"

making it easier for firms to use their in-house research facilities to acquire technology.

Measuring the Growth of Industrial Research

Although recent historiography on U.S. industrial research has focused primarily on the electrical industry (an exception is Hounshell and Smith 1988), the limited data on the growth of industrial research activity suggest that chemicals and related industries were the dominant early investors in R&D. The chemicals, glass, rubber, and petroleum industries accounted for nearly 40% of the number of laboratories founded between 1899 and 1946. The chemicals sector also dominated research employment between 1921 and 1946. In 1921, the chemicals, petroleum, and rubber industries accounted for slightly more than 40% of total research scientists and engineers in manufacturing. The dominance of chemicals-related industries as research employers was supplemented during the period by industries whose product and process technologies drew heavily on physics. Electrical machinery and instruments accounted for less than 10% of total research employment in 1921. By 1946, however, these two industries contained more than 20% of all scientists and engineers employed in industrial research in U.S. manufacturing, and the chemicals-based industries had increased their share to slightly more than 43% of total research employment.

Table 1 provides data on research laboratory employment for 1921, 1927, 1933, 1940, and 1946 in nineteen manufacturing industries and in manufacturing overall (excluding miscellaneous manufacturing industries). Employment of scientists and engineers in research within these manufacturing industries grew from roughly 3,000 in 1921 to nearly 46,000 by 1946.[13] The ordering of

---

[13] The data in Table 1 were drawn originally from the National Research Council surveys of industrial research employment, as tabulated in Mowery (1981).

Table 1. Employment of Scientists and Engineers in Industrial Research Laboratories in U.S. Manufacturing Firms, 1921–46.

Industry	Number Employed (Employment per 1000 Production Workers)				
	1921	1927	1933	1940	1946
Food/beverages	116	354	651	1712	2510
	(.19)	(.53)	(.973)	(2.13)	(2.26)
Tobacco	—	4	17	54	67
		(.031)	(.19)	(.61)	(.65)
Textiles	15	79	149	254	434
	(.015)	(.07)	(.15)	(.23)	(.38)
Apparel	—	—	—	4	25
				(.005)	(.03)
Lumber products	30	50	65	128	187
	(.043)	(.16)	(.22)	(.30)	(.31)
Furniture	—	—	5	19	19
			(.041)	(.10)	(.07)
Paper	89	189	302	752	770
	(.49)	(.87)	(1.54)	(2.79)	(1.96)
Publishing	—	—	4	9	28
			(.015)	(.03)	(.06)
Chemicals	1102	1812	3255	7675	14,066
	(5.2)	(6.52)	(12.81)	(27.81)	(30.31)
Petroleum	159	465	994	2849	4750
	(1.83)	(4.65)	(11.04)	(26.38)	(28.79)
Rubber products	207	361	564	1000	1069
	(2.04)	(2.56)	(5.65)	(8.35)	(5.2)
Leather	25	35	67	68	86
	(.09)	(.11)	(.24)	(.21)	(.25)

(cont.)

Table 1. (cont.)

Industry	Number Employed (Employment per 1000 Production Workers)				
	1921	1927	1933	1940	1946
Stone/clay/glass	96 (.38)	410 (1.18)	569 (3.25)	1334 (5.0)	1508 (3.72)
Primary metals	297 (.78)	538 (.93)	850 (2.0)	2113 (3.13)	2460 (2.39)
Fabricated metal products	103 (.27)	334 (.63)	500 (.153)	1332 (2.95)	1489 (1.81)
Nonelectrical machinery	127 (.25)	421 (.65)	629 (1.68)	2122 (3.96)	2743 (2.2)
Electrical machinery	199 (1.11)	732 (2.86)	1322 (8.06)	3269 (13.18)	6993 (11.01)
Transportation equipment	83 (.204)	256 (.52)	394 (1.28)	1765 (3.24)	4491 (4.58)
Instruments	127 (.396)	234 (.63)	581 (2.69)	1318 (4.04)	2246 (3.81)
TOTAL	2775	6274	10,918	27,777	45,941

Source: Mowery (1981).

industries by research intensity is remarkably stable – chemicals, rubber, petroleum, and electrical machinery are among the most research-intensive industries, accounting for 48% to 58% of total employment of scientists and engineers in industrial research within manufacturing, throughout this period. The major prewar research employers remained among the most research-intensive industries well into the postwar period despite the growth in federal funding for research in industry. Chemicals, rubber, petroleum, and electrical machinery accounted for more than 53% of industrial research employment in manufacturing in 1940 and represented 39.7% of research employment in U.S. manufacturing in 1995 (National Science Foundation 1996).[14]

Industrial Research and the Universities, 1900–40

The pursuit of research was recognized as an important professional activity within both U.S. industry and higher education only in the late 19th century, and research in both venues was influenced by the example (and in the case of U.S. industry, by the competitive pressure) of German industry and academia. The reliance of many U.S. universities on state government funding, the modest scope of this funding, and the rapid expansion of their training activities all supported the growth of formal and informal linkages between industry and university research. U.S. universities formed a focal point for the external technology-monitoring activities of many U.S. industrial research laboratories before 1940, and at least some of these university-industry linkages involved the development and commercialization of new technologies and products.

Linkages between academic and industrial research were powerfully influenced by the decentralized structure and funding of U.S. higher education, especially the public institutions within the system. Public funding created a U.S. higher education system that was substantially larger than that of most European nations. The

---

[14] Similar stability is revealed in the geographic concentration of industrial research employment during this period. Five states (New York, New Jersey, Pennsylvania, Ohio, and Illinois) contained more than 70% of the professionals employed in industrial research in 1921 and 1927; their share declined modestly, to slightly more than 60%, by 1940 and 1946. In other words, the regional concentration of high-technology firms and R&D activities within the United States was well established prior to the 1950s.

source of this public funding, however, was equally important. The prominent role of state governments in financing the prewar U.S. higher education system led public universities to seek to provide economic benefits to their regions through formal and informal links to industry (Rosenberg and Nelson 1994).

Both the curriculum and research within U.S. higher education were more closely geared to commercial opportunities than was true in many European systems of higher education. Swann (1988) describes the extensive relationships between academic researchers, in both public and private educational institutions, and U.S. ethical drug firms that developed after World War I.[15] Hounshell and Smith (1988, pp. 290–292) document a similar trend for the Du Pont Company, which funded graduate fellowships at 25 universities during the 1920s and expanded its program during the 1930s to include support for postdoctoral researchers. During the 1920s, colleges and universities to which the firm provided funds for graduate research fellowships also asked Du Pont for suggestions for research, and in 1938 a leading Du Pont researcher left the firm to head the chemical engineering department at the University of Delaware (Hounshell and Smith 1988, p. 295).

Still another university with strong ties with local and national firms was MIT, founded in 1862 with Morrill Act funds by the state of Massachusetts.[16] In 1906, MIT's electrical engineering

---

[15] According to Swann (1988, p. 50), Squibb's support of university research fellowships expanded (in current dollars) from $18,400 in 1925 to more than $48,000 in 1930 and accounted for one seventh of the firm's total R&D budget for the period. By 1943, according to Swann, university research fellowships amounting to more than $87,000 accounted for 11% of Eli Lilly and Company's R&D budget. Swann cites similarly ambitious university research programs sponsored by Merck and Upjohn.

[16] The MIT example also illustrates the effects of reductions in state funding on universities' eagerness to seek out industrial research sponsors. Wildes and Lindgren (1985, p. 63) note that the 1919 withdrawal by the Massachusetts state legislature of financial support for MIT, along with the termination of the

department established an advisory committee that included Elihu Thomson of General Electric, Charles Edgar of the Edison Electric Illuminating Company of Boston, Hammond V. Hayes of AT&T, Louis Ferguson of the Chicago Edison Company, and Charles Scott of Westinghouse (Wildes and Lindgren, 1985, pp. 42–43). The department's Division of Electrical Engineering Research, established in 1913, received regular contributions from General Electric, AT&T, and Stone and Webster, among other firms. MIT was later to play an important role in the development of U.S. chemical engineering and worked closely with U.S. chemicals and petroleum firms in this effort (see further discussion subsequently).

Training by public universities of scientists and engineers for employment in industrial research also linked U.S. universities and industry during this period. The PhDs trained in public universities were important participants in the expansion of industrial research employment during this period (Thackray 1982, p. 211).[17] The size of this trained manpower pool was as important as its quality; although the situation was improving in the decade before 1940, Cohen (1976) noted that virtually all "serious" U.S. scientists completed their studies at European universities. Thackray et al. (1985) argue that American chemistry research during this period attracted

Institute's agreement with Harvard University to teach Harvard engineering courses, led MIT President Richard C. Maclaurin to establish the Division of Industrial Cooperation and Research. This organization was financed by industrial firms in exchange for access to MIT libraries, laboratories, and staff for consultation on industrial problems. Still another institutional link between MIT and a research-intensive U.S. industry, the Institute's School of Chemical Engineering Practice, was established in 1916 (Mattill 1991).

[17] Hounshell and Smith (1988, p. 298) report that 46 of the 176 PhDs overseen by Carl Marvel, longtime professor in the University of Illinois chemistry department, went to work for one firm, Du Pont. According to Thackray (1982, p. 221), 65% of the 184 PhDs overseen by Professor Roger Adams of the University of Illinois during 1918–58 went directly into industrial employment. In 1940, 30 of the 46 PhDs produced by the University of Illinois chemistry department were first employed in industry.

attention (in the form of citations in other scientific papers) as much because of its quantity as its quality.[18]

The Federal Role in U.S. R&D Before 1940

In spite of the permissive implications of the "general welfare" clause of the U.S. Constitution, federal support for science prior to World War II was limited. During World War I, the military operated the R&D and production facilities for the war effort, with the exception of the munitions industry, where the federal government relied on DuPont.[19] If one of the armed services identified a scientific need, a person with the appropriate qualifications was drafted into

---

[18] "[F]rom comparative obscurity before World War I, American chemistry rose steadily in esteem to a position of international dominance. Almost half of the citations in the *Annual Reports* [*Annual Reports in Chemistry*, described as 'a central British review journal'] in 1975 were to American publications. Similarly, almost half the citations to non-German-language literature in *Chemische Berichte* [the 'central German chemical journal'] in 1975 went to American work. It is striking that this hegemony is the culmination of a fifty-year trend of increasing presence, and not merely the result of post–World War II developments. Second, it is clear that the increasing attention received in the two decades before World War II reflected the growing *volume* of American chemistry, rather than a changed assessment of its worth. Since World War II, however, in both *Chemische Berichte* and the *Annual Reports*, American chemistry has been cited proportionately more than is warranted by increasing quantity alone. The prominence of American work within the international literature has been sustained by quality" (Thackray et al. 1985, p. 157; emphasis in original).

[19] Sapolsky (1990, p. 13), describing the organization of the World War II Office of Scientific Research and Development, characterized the experience of World War I as follows: "[T]he military was initially reluctant to admit a need for outside assistance in the design of weapons, and then insisted on dominating the hurriedly created scientific effort that only began with the involvement of American troops in the fighting. Scientists who wished to contribute to the war by doing weapons-related research were required, with rare exception, to accept military commissions and to work at government facilities under military command procedures. Research priorities were determined by the military, and no attention was paid to linking weapon development to operational experience."

that branch. One legacy of wartime programs for technology development was the National Advisory Committee on Aeronautics (NACA), founded in 1915 to "investigate the scientific problems involved in flight and to give advice to the military air services and other aviation services of the government" (Ames 1925). The Committee, which was absorbed by the National Aeronautics and Space Administration in 1958, made important contributions to the development of new aeronautics technologies for both civilian and military applications throughout its existence, but was particularly important during the pre-1940 era.[20]

For 1940, the last year that was not dominated by the vast expenditures associated with wartime mobilization, total federal expenditures for research, development, and R&D plant amounted to $74.1 million (National Science Foundation 1966, p. 149). Of that, Department of Agriculture expenditures amounted to $29.1 million, or 39%. In 1940, the Department of Agriculture's research budget exceeded that of the agencies that would eventually be combined in the Department of Defense, whose total research budget amounted to $26.4 million. Between them, these agencies accounted for 75% of all federal R&D expenditures. The claimants on the remaining 25%, in descending order of importance, were the Department of the Interior ($7.9 million), the Department of Commerce ($3.3 million), the Public Health Service ($2.8 million), and the National Advisory Committee on Aeronautics ($2.2 million).

Federal expenditures for R&D throughout the 1930s constituted 12% to 20% of total U.S. R&D expenditures. Industry accounted for about two thirds of the total. The remainder came from

[20] Vannevar Bush, who chaired the Advisory Committee during the 1930s, cited NACA approvingly as a model for his postwar proposal of a National Research Foundation in his influential 1945 report, *Science: The Endless Frontier*: "The very successful pattern of organization of the National Advisory Committee for Aeronautics, which has promoted basic research on problems of flight during the past thirty years, has been carefully considered in proposing the method of appointment of Members of the Foundation and in defining their responsibilities" (p. 40).

universities, state governments, private foundations, and research institutes. One estimate suggests that state funds may have accounted for as much as 14% of university research funding during 1935 and 1936 (National Resources Planning Board 1942, p. 178). Moreover, the contribution of state governments to nonagricultural university research appears from these data to have exceeded the federal contribution, in sharp contrast to the postwar period.

The Impact of World War II on the Structure of U.S. R&D

War preparations and the entry of the United States into World War II in December 1941 transformed the bucolic picture of federal R&D expenditures discussed above. Funding for the nondefense categories of prewar R&D declined substantially in real terms during the war. But overall federal R&D expenditures (in 1930 dollars) soared from $83.2 million in 1940 to a peak of $1313.6 million in 1945. Over the same period, the research expenditures of the Department of Defense rose from $29.6 million to $423.6 million (in 1930 dollars).

The success and the organizational structure of the massive federal wartime R&D program yielded several important legacies. The successful completion of the Manhattan Project, whose research budget in the peak years 1944 and 1945 substantially exceeded that of the Department of Defense, created a research and weapons production complex that ushered in the age of truly "big science." Ironically, the Manhattan Project's success in creating weapons of unprecedented destructive power contributed to rosy postwar perceptions of the constructive possibilities of large-scale science for the advance of societal welfare.[21]

---

[21] Some of the large R&D programs that were mounted under the exigencies of war did, however, generate huge societal benefits in the postwar years. A "crash" wartime program made penicillin, perhaps the greatest medical breakthrough

Far smaller in financial terms, but significant as an institutional innovation, was the Office of Scientific Research and Development (OSRD), a civilian agency directed by Vannevar Bush. OSRD entered into research contracts with private firms and universities – the largest single recipient of OSRD grants and contracts during wartime (and the inventor of that device beloved of university research administrators, institutional overhead) was MIT, with 75 contracts for a total of more than $116 million (Owens, 1994; Gruber, 1995). The largest corporate recipient of OSRD funds, Western Electric, accounted for only $17 million (Pursell 1977, p. 364).

The contrast between the organization of wartime R&D in the First and Second World Wars reflects the more advanced university and private sector research capabilities during the second global conflict. The contractual arrangements developed by OSRD during the Second World War allowed the Office to tap the broad array of academic and industrial R&D capabilities that had developed during the interwar period. Members of the scientific community were called upon to recommend and to guide as well as to participate in scientific research with military payoffs. OSRD was not subordinated to the military and had direct access to the President and to the pertinent congressional appropriations committees.

The success of these wartime contractual arrangements with the private sector contributed to the growth of a postwar R&D system that relied heavily on federal financing of extramural research and development.[22] In 1940, the bulk of federal R&D went to support

---

of the 20th century, widely available for the treatment of infectious diseases. Another large-scale program (discussed in a subsequent section) made low-cost synthetic rubber widely available and had lasting effects on the U.S. chemicals and petrochemicals industries, and wartime research in microelectronics, directed toward military goals such as improvement of radar systems, provided a rich legacy of enlarged technological capabilities to the postwar world.

[22] For a careful, although now somewhat dated, treatment of these contractual issues, see Danhof (1968).

research performed within the public sector – by federal civil servants, as in the National Bureau of Standards, the Department of Agriculture, and the Public Health Service, or by state institutions financed by federal grants, as in the agricultural experiment stations. In the postwar period, by contrast, most federal R&D funds have supported the performance of research by nongovernmental organizations. Moreover, the dramatic growth in federal funding for research in universities contributed to the creation of a huge basic research complex in this sector. Combined with large federal procurement contracts, federal funding for R&D in industry had profound consequences for the emergence of a series of new, high-technology industries in the postwar period.

The Postwar Structure, 1945–95

Introduction

Two salient features of postwar R&D spending are the magnitude of the overall national R&D investment and the size of the federal R&D budget. Throughout the period from 1940 through 1995, federal R&D spending was a large fraction of a very large national R&D investment. The total volume of resources devoted to R&D since the end of the Second World War is not only large by comparison with our earlier history, but also by comparison with other Organization for Economic Cooperation and Development member countries. Indeed, as late as 1969, when the combined R&D expenditures of the largest foreign industrial economies (West Germany, France, the United Kingdom, and Japan) were $11.3 billion, those for the United States were $25.6 billion. Not until the late 1970s did the combined total for those four countries exceed that of the United States.

Within the postwar R&D system, federal expenditures have financed somewhere between one half and two thirds of total R&D,

the great bulk of which is performed by private industry. In 1995, industry performed 71% of total national R&D; slightly more than 36% of federally funded R&D was performed in private industry. Only 27% of federally financed R&D was performed in federal intramural laboratories, although federal sources financed more than 35% of all U.S. R&D in 1995. Federal funds have been especially important in supporting basic research. Federal sources financed 58% of all U.S. basic research in 1995, although federal research establishments perform only 9.1% of U.S. basic research. Universities have increased in importance as basic research performers during this period. In 1953, less than one third of all basic research was performed in universities and Federally Funded R&D Centers (FFRDCs) at universities and colleges. In 1996, however, these institutions performed 61% of all U.S. basic research (all figures from National Science Foundation [1996]).

Perhaps the other most significant feature of the postwar U.S. R&D system is the extent to which the federal presence within it assumed a shape that differed dramatically from that envisioned by one of the most famous and influential figures in U.S. science policy during this century, Vannevar Bush. In response to a request from President Roosevelt (a request that he had solicited), Bush, the overseer of wartime R&D policy, drafted the famous 1945 report on postwar federal science policy, *Science: The Endless Frontier*. Anticipating subsequent economic analysis, Bush argued that basic research was the ultimate source of economic growth. He advocated the creation of a single federal agency charged with responsibility for funding basic research in all defense and nondefense areas, including health. Bush's advocacy of civilian direction of basic military research reflected his wartime experiences, as did his recommendation that his "National Research Foundation" focus its financial support on extramural research, primarily within the nation's universities (see Mowery [1997], for a more detailed discussion). The complexities of postwar domestic politics, as well as Bush's resistance to Congressional oversight of his proposed agency, ultimately doomed his proposal. Rather than a single, civilian agency overseeing all of federal science policy and funding, various mission agencies, including the military and the National Institutes of Health, assumed major roles in supporting basic and applied research. By the end of fiscal 1950, more than 90% of federal R&D spending was controlled by the Defense Department and the Atomic Energy Agency.

Defense-Related R&D and Procurement

The military services have dominated the federal R&D budget for the past 30 years, falling below 50% of federal R&D obligations in only three years (see Fig. 1). In 1960 defense research constituted no less than 80% of federal R&D funds. It declined sharply from that level (a decline offset by the growth of the space program) and hovered around the 50% level until the early 1980s, when it rose swiftly again, and began a slow decline in the late 1980s and early 1990s.

As a result of the development emphasis in defense R&D and the large size of the defense R&D budget, the distribution of the federal R&D budget across industry sectors is highly concentrated. Nearly 55% of all federal R&D in 1993 went to just two industries – aircraft and missiles (42%), and scientific and measurement instruments (12%) (National Science Board 1996). A number of analysts have argued that military expenditures strengthened the commercial innovative capabilities of U.S. firms in these industries during the postwar period (Derian 1990). Assessing the commercial impact of postwar U.S. military R&D spending, however, is complicated by the fact that the influence of such spending is confounded with that of military procurement. Separating the influence of Pentagon R&D spending from that of Pentagon procurement is further complicated by the practice of paying a percentage of military procurement contracts to defense suppliers as an "independent R&D"

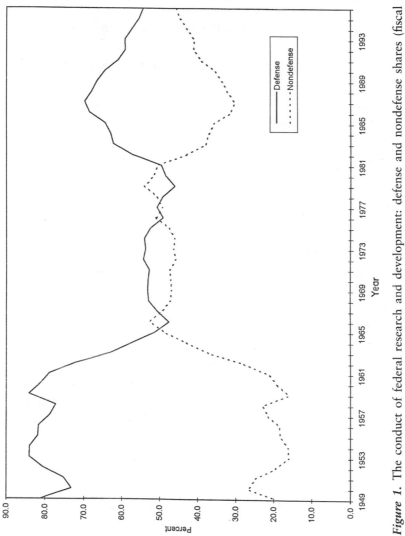

Figure 1. The conduct of federal research and development: defense and nondefense shares (fiscal years 1949–96). Outlays for 1995 are estimated. Outlays for 1996 are proposed. *Source:* U.S. Office of Management and Budget (1995).

33

allowance that was generally not included in either the formal defense R&D budget or the reported R&D expenditures of recipient firms.

In some key technologies, such as aircraft, semiconductors, and computers, defense-related R&D investments during the 1950s generated important technological "spillovers" from military to civilian applications. Among the most important influences on these technological spillovers was the extent of generic similarity of civilian and military requirements for a technology (see Nelson [1984], for further discussion). Frequently, commercial and military requirements for performance, cost, ruggedness, and so forth closely resemble one another early in the development of a new technology.[23] This broad similarity in requirements appears to have been associated with significant spillovers in microelectronics in the early 1960s, when the demands of the commercial and military markets for miniaturization, low heat in operation, and ruggedness did not diverge too dramatically. The similarity of military and commercial requirements in microelectronics declined, however, and military demand now accounts for a much smaller share of total U.S. semiconductor output. During the 1950s and 1960s, the jet engine was applied in military strategic bombers, transports, and tankers, all of which had fuselage design and engine performance requirements that resembled some of those for commercial air transports. Over time, however, the size and even the direction of spillovers in these technologies appears to have changed, as we have noted.[24]

---

[23] Interestingly, computer software appears to be one technology in which military and commercial demands diverged significantly from the inception of this industry. As a result, such military-civilian spillovers as occurred in this technology flowed from defense-related funding of academic research in computer science (Langlois and Mowery 1996).

[24] Flamm and McNaugher (1989) suggest that changes in Defense Department R&D policy have contributed to declining military-civilian technology spillovers. They cite declines in the share of basic research in Defense Department R&D spending, as well as increased Congressional demands that these

Changing Roles for University Research

Another new element in the structure of the postwar U.S. research system was the expansion of publicly funded research in U.S. universities. Although Bush's recommendations of a single federal funding agency for basic research were not implemented, and his advocacy of institutional, rather than project, funding, also was ignored, U.S. universities nonetheless enjoyed significant increases in federal R&D support during this period. By any measure, academic research grew dramatically. From an estimated level of nearly $500 million in 1935–36, university research (excluding FFRDCs) grew to more than $2.4 billion in 1960 and $16.8 billion in 1995 (National Science Board 1996; all amounts in 1987 dollars). The increase in federal support of university research transformed major U.S. universities into worldwide centers for the performance of scientific research, a role that differs significantly from that of U.S. academia in the prewar years.

The federal government did not confine itself to expanding the demand for university research. Federal actions on the supply side enlarged the pool of scientific personnel and supported the acquisition of the physical equipment and facilities essential to the performance of high-quality research. In the case of computer science, federal support for university purchases of large mainframe computers were indispensable to the institutionalization of a new academic discipline in U.S. universities. After the Second World War, federal programs increased financial aid for students in higher education.[25]

R&D programs yield near-term applications in weapons systems, as two factors that have reduced such spillovers.

[25] The best known of these was the GI Bill, which provided substantial financial support to all veterans who enrolled in college-level educational programs; others include graduate fellowships supported by National Science Foundation (NSF) and Atomic Energy Commission (AEC) funds, training fellowships from the National Institutes of Health, and the National Defense Education Act fellowships.

By simultaneously providing funds for university education and for the support of university research, the federal government strengthened the university commitment to research and reinforced the link between research and teaching. The combination of research and teaching in higher education has been carried much further in the United States than elsewhere. In Europe and Japan, for example, a larger fraction of research is carried out in specialized research institutes not connected directly with higher education and in government-operated laboratories.[26]

Since the early 1980s, the central role of the federal government in supporting academic research has been supplemented by increased funding from industry, and university-industry research

[26] See National Research Council (1982), Okimoto and Saxonhouse (1987). Sharp (1989, pp. 12–13) argues that the less prominent role played in scientific research by European universities has contributed to the slower growth of small biotechnology firms: "A researcher at a Centre Nationale de Recherche Scientifique (CNRS) laboratory in France, or at a Max Planck Institute laboratory in Germany, is the full time employee of that institution. As such his/her prime responsibility is to public, not private science. Moreover, as a full time employee, he/she will not find it easy to undertake the 'mix' of research frequently undertaken by an American professor, who combines an academic post with consultancy in the private sector. Indeed the tradition of funding US academic posts for only nine months of the year, expecting the academic who wishes to carry out research in the summer to raise research funds to meet the remaining three months of salary, explicitly encourages the entrepreneurial academic. In stark contrast, his/her German opposite number at a Max Planck Institute will find all research costs, including staff and equipment, met as part of institutional overheads. The opportunity cost of leaving such a research environment for the insecurity of the small firm is all the greater since, once off the academic ladder in West Germany, it is more difficult to climb back on again. The same goes for the opposite number in France, and with the additional disincentive that French researchers are civil servants and dropping out of the system means *both* losing security of tenure/accumulated benefits *and* difficulty in re-entry should the need arise. In the circumstances, it is not perhaps so surprising that few spin-offs from public sector research arise, nor, for that matter, that in Europe most such spin-offs are to be found in the UK, where the organisation of academic science most closely matches that of the US. In the UK, it is notable that – with the exception of Celltech and the Agricultural Genetics Company – most of the spin-offs from biotechnology have come from the universities."

linkages have attracted considerable comment. As our discussion of the pre-1940 U.S. R&D system notes, however, these linkages were well-established before World War II. Indeed, the share of university research expenditures financed by industry appears to have declined through much of the postwar period.[27] In 1953, industry financed 11% of university research, a share that declined to 5.5% in 1960 and 2.7% in 1978, in part as a result of increased federal government funding of academic research. By 1992, industrial support for university research had rebounded to account for roughly 7% of academic research spending, and industry funding has remained at roughly this level through 1997.

Another important academic research institution that has received less attention from scholars, yet has proven to be a source of enormously significant innovations during the postwar period, is the academic medical center. By combining scientific research with clinical practice, the academic medical center in the United States has been able to link science and innovation to a remarkable

---

[27] One might instead argue that the weakening of university-industry research linkages during a significant portion of the postwar period was the real departure from historical trends. Hounshell and Smith (1988, p. 355) summarize a 1945 memo from Elmer Bolton, director of what was to become the Du Pont Company's central research laboratory, that made a case for greater self-reliance by the firm in its basic research: "Three things were necessary: Du Pont had to strengthen its research organizations and house them in modern research facilities; the company's existing processes had to be improved and new processes and products developed; and 'fundamental research, which will serve as a background for new advances in applied chemistry, should be expanded not only in the Chemical Department but should [also] be increased in our industrial research laboratories and the Engineering Department.' Bolton stressed that it was no longer 'possible to rely to the same extent as in the past upon university research to supply this background so that in future years it will be necessary for the Company to provide this knowledge to a far greater extent through its own efforts.' To 'retain its leadership' Du Pont had 'to undertake on a much broader scale fundamental research in order to provide more knowledge to serve as a basis for applied research.' " Swann (1988, pp. 170–181) also argues that research links between U.S. universities and the pharmaceuticals industry weakened significantly in the immediate aftermath of World War II, in part as a result of vastly increased federal research funding for academic research in the health sciences.

degree, making possible the rapid collection by scientists of feedback from practitioners in the development of new medical devices and procedures, facilitating clinical tests of new pharmaceuticals, and contributing powerfully to innovations in both pharmaceuticals and medical devices. The combination of science and clinical applications in one institution is unusual – as Henderson, Orsenigo, and Pisano (1998) and Gelijns and Rosenberg (1998) point out, most Western European medical institutions emphasize clinical practice and applications much more heavily than scientific research. As is true of other components of the U.S. academic research enterprise, the remarkable advances in biomedical technologies during the postwar period and the large, multifunction academic medical centers have benefited from large infusions of federal funding. The National Institutes of Health in particular have enjoyed strong bipartisan political support for decades, and R&D funding for biomedical science and applications has been abundant. But this discussion also underscores the extent to which U.S. universities and academic research facilities have maintained an important presence in the "D," as well as the "R," of R&D throughout the postwar period.

Research in Industry

As the discussion in the previous section makes clear, private industry continued to dominate U.S. R&D during the postwar period amid shifts in the sources of its R&D funding. In 1993, although it performed 68% of total U.S. research and development, industry accounted for slightly more than 50% of total U.S. R&D investment. Its primacy as a performer of R&D, however, meant continued growth in employment within industrial research – from less than 50,000 in 1946 (see Table 1) to roughly 300,000 scientists and engineers in 1962, 376,000 in 1970, and almost 800,000 in 1996 (Birr 1966; U.S. Bureau of the Census 1987, p. 570; National Science Foundation 1996, p. 109).

U.S. antitrust policy remained an important influence on indus-trial research and innovation during the postwar period, but both the policy and the nature of its influence changed. The appoint-ment of Thurman Arnold in 1938 to head the Antitrust Division of the U.S. Justice Department, combined with growing criticism of large firms and economic concentration (e.g., the investigations of the federal Temporary National Economic Committee), produced a much tougher antitrust policy that extended well into the 1970s. The cases filed by Arnold and his successors, many of which were decided or resolved through consent decrees in the 1940s and early 1950s, transformed the postwar industrial research strategies of many large U.S. firms.

This revised antitrust policy made it more difficult for large U.S. firms to acquire firms in "related" technologies or industries[28] and led them to rely more heavily on intrafirm sources for new tech-nologies. In the case of Du Pont, the use of the central laboratory and development department to seek technologies from external sources was ruled out by senior management as a result of the per-ceived antitrust restrictions on acquisitions in related industries. As a result, internal discovery (as well as development) of new prod-ucts became paramount (Hounshell and Smith [1988] emphasize the firm's postwar expansion in R&D and its search for "new nylons"[29]), in contrast to the firm's R&D strategy before World War II.

---

[28] Hawley (1966) analyzes the shifting antitrust policies of the New Deal. Arnold took office in 1938 and during 1938–42 filed 312 antitrust cases, considerably above the 46 filed during 1932–37 or the 70 filed during 1926–31 (Fligstein 1990, p. 168).

[29] Hounshell and Smith (1988) and Mueller (1962) both argue that the discov-ery and development of nylon, one of Du Pont's most commercially successful innovations, was in fact atypical of the firm's pre-1940 R&D strategy. Rather than being developed to the point of commercialization following its acquisi-tion by Du Pont, nylon was based on the basic research of Carothers within Du Pont's central corporate research facilities. The successful development of nylon from basic research through to commercialization nevertheless exerted a strong

In Du Pont's case, this shift in R&D strategy weakened the links between the firm's growing central research facilities, which increasingly concentrated their efforts on basic research, and its operating divisions. The R&D efforts of the established business units focused on increasingly costly improvements in existing processes and products, and the overall productivity of Du Pont R&D suffered (Hounshell and Smith 1988, p. 598). The inward focus of Du Pont research appears to have impaired the firm's postwar innovative performance, even as its central corporate research laboratory gained a sterling reputation within the global scientific community.

In other U.S. firms, senior managers sought to maintain growth through the acquisition of firms in unrelated lines of business, creating conglomerate firms with few if any technological links among products or processes. Chandler (1990) and others (e.g., Ravenscraft and Scherer 1987; Fligstein 1990) have argued that this extensive diversification weakened senior management understanding of and commitment to the development of the technologies that historically had been essential to competitive success, eroding the quality and consistency of decisionmaking on technology-related issues.

RCA, for example, followed a conglomerate diversification strategy but maintained its large fundamental research "campus" near Princeton, New Jersey, that made important contributions to military and consumer electronics technologies. The firm encountered growing difficulties, however, in reaping the commercial returns to its considerable research capabilities. Thus, RCA pursued development of the hugely expensive and unsuccessful videodisc home-entertainment technology (Graham 1986a) but failed to maintain its dominant position in color television receivers.

---

influence on Du Pont's postwar R&D strategy, not least because of the fact that many senior Du Pont executives had direct experience with the nylon project. Hounshell (1992) argues that Du Pont had far less success in employing the "lessons of nylon" to manage such costly postwar synthetic fiber innovations as Delrin.

At the same time that established firms were shifting the R&D strategies that many had employed since the early 20th century, new firms began to play an important role in the development of the technologies spawned by the postwar U.S. R&D system. The prominence of small firms in commercializing new electronics technologies in the postwar United States contrasts with their more modest role in this industry during the interwar period. In industries that effectively did not exist before 1940, such as computers and biotechnology, major innovations were commercialized largely through the efforts of new firms.[30] These postwar U.S. industries differ from their counterparts in Japan and most Western European economies, in which established electronics and pharmaceuticals firms dominated the commercialization of these technologies.

The arguments of Chandler (1990) and Pavitt (1990) about large firms' dominance of new technologies thus require some qualification if applied to high-technology industries in the postwar United States. The significant technological contributions made by large firms in semiconductors, for example, were not matched by their role in commercializing these technologies. In biotechnology, small firms have played an important role in expanding the technology pool and in commercializing its contents. Moreover, in both semiconductors and computers, new small firms grew rapidly to positions of considerable size and market share.

---

[30] This is not to deny the major role played by such large firms as IBM in computers and AT&T in microelectronics. In other instances, large firms have acquired smaller enterprises and applied their production or marketing expertise to expand markets for a new product technology. Nonetheless, it seems apparent that start-up firms have been far more active in commercializing new technologies in the United States than in other industrial economies. Malerba (1985) and Tilton (1971) stress the importance of new, small firms in the U.S. semiconductor industry; Flamm (1988) describes their significant role in computer technology; and Orsenigo (1988) and Pisano, Shan, and Teece (1988) discuss the importance of these firms in the U.S. biotechnology industry. Bollinger et al. (1983) survey some of the literature on the "new technology-based firm."

Several factors contributed to this prominent role of new, small firms in the postwar U.S. innovation system. The large basic research establishments in universities, government, and a number of private firms served as important "incubators" for the development of innovations that "walked out the door" with individuals who established firms to commercialize them. This pattern was particularly significant in the biotechnology, microelectronics, and computer industries. Indeed, high levels of labor mobility within regional agglomerations of high-technology firms served both as an important channel for technology diffusion and as a magnet for other firms in related industries.

The foundation and survival of vigorous new firms also depended on a sophisticated private financial system that supported new firms during their infancy. The U.S. venture capital market played an especially important role in the establishment of many microelectronics firms during the 1950s and 1960s and has contributed to the growth of the biotechnology and computer industries. According to the Office of Technology Assessment (1984, p. 274), the annual flow of venture capital into industrial investments ranged between $2.5 and $3 billion between 1969 and 1977. Venture capital–supported investments directed specifically to new firms, however, were substantially smaller, averaging roughly $500 million annually during the 1980s (Florida and Smith 1990). Investment funds from venture capital were gradually supplemented by public equity offerings.[31]

---

[31] See Perry (1986), Mowery and Steinmueller (1994). Sharp (1989, pp. 9–10) argues that "the venture capital market in Europe is underdeveloped. The most active venture capital market is in the UK where some half dozen funds specialising in investment in biotechnology are active and an estimated total of over $1 billion invested since 1980... The doyen of this market is the Rothschild Fund Biotechnology Investments Ltd (BIL) – now capitalised at $200 million and the largest specialist fund in Europe. By contrast, the largest German venture capital fund, Techno Venture Management, established in 1984, had an initial capitalisation of $10 million and in 1989 is worth only $50 million. The availability of venture capital, however, is only one part of the equation. BIL,

Western European economies have yet to spawn similarly abundant sources of risk capital for new enterprises in high-technology industries (Sharp 1989, pp. 9–10). Okimoto (1986, p. 562) estimated that Japanese venture capital firms provided no more than $100 million in financing in 1986.

Formal intellectual property protection had complex effects on the postwar growth of new firms in a number of U.S. high-technology industries. In several of these industries, relatively weak formal protection aided the early growth of new firms. Commercialization of microelectronics and computer hardware and software innovations by new firms was aided by a permissive intellectual property regime that facilitated technology diffusion and reduced the burden on young firms of litigation over inventions that originated in part within established firms. In microelectronics and computers, liberal licensing and cross-licensing policies were byproducts of antitrust litigation, illustrating the tight links between these strands of U.S. government policy. The 1956 consent decrees that settled federal antitrust suits against IBM and AT&T both mandated liberal licensing of these technologies, lowering barriers to entry by new firms into the embryonic computer and semiconductor industries (Flamm 1988).

The entry and growth of new high-technology firms benefited from yet another postwar federal policy, military procurement. As our discussion of semiconductors noted, military procurement policies contributed to the growth of a number of new firms in microelectronics and contributed to high levels of technology spillover among these firms. Especially in the early days of this industry,

for example, whose investments span biotechnology and medical technology, have not found in Europe the quality of investment they are seeking. 75 per cent of their investments are in the US, only 25 per cent in Europe, and these concentrated almost entirely in the UK. This pattern of investment is mirrored by nearly all the investment funds, all of which invest a large proportion of their investments in biotechnology in the small firm sector in the US, and only a very small proportion in small firms in Europe."

the commercial consequences of military procurement policies were heightened further by the technological spillovers that flowed from military to civilian products.

Conclusion

Our discussion of the evolution of the U.S. R&D system has emphasized two structural transformations – that associated with the emergence of large corporate enterprises at the turn of the century, many of which pioneered in the development of the industrial research laboratory, and the changes wrought in this system by World War II and its aftermath. The structure (if not necessarily the scale) of the pre-1940 U.S. R&D system resembled those of other leading industrial economies of the era, such as the United Kingdom, Germany, and France – industry was a significant funder and performer of R&D and central government funding of R&D was modest. By contrast, the postwar U.S. R&D system differed from those of other industrial economies in at least three aspects: small, new firms were important entities in the commercialization of new technologies; defense-related R&D funding and procurement exercised a pervasive influence in the high-technology sectors of the US economy; and US antitrust policy during the postwar period was unusually stringent.

These three characteristics of the postwar system were mutually interdependent. Defense-related R&D and procurement were indispensable to the growth of start-up firms in the semiconductor and computer industries. Antitrust policies contributed to the rapid diffusion of intellectual property throughout the nascent computer and semiconductor industries. In addition, of course, the support by federal agencies of research in firms and universities contributed to an abundant supply of new developments with commercial potential. The commercialization of these developments often relied on the extension to much smaller firms of the equity-based system of industrial finance that distinguishes the U.S. economy from those of Germany and Japan. Many of the technologies developed with the support of defense-related R&D spending during the 1950s and 1960s also found profitable and substantial applications in commercial markets.

The factors that gave rise to the unique structural characteristics of the US postwar R&D system may lose much of their influence in the years ahead. First and most obviously, the large defense budgets that characterized the U.S. economy during the Cold War have been reduced by nearly one third and are likely to remain below the real spending levels of the 1980s for the foreseeable future. In addition to its declining magnitude, the commercial technology spillovers associated with high levels of defense-related R&D spending also appear to have declined since the 1950s and 1960s, as we note in this chapter. Lower levels of defense-related spending on R&D and procurement, combined with lower levels of technological spillovers, will reduce the influence of defense spending on commercial innovation in the United States.

Antitrust policy has also undergone considerable change since 1980. The Reagan administration substantially shifted its enforcement philosophy away from the tough approach adopted by virtually all earlier postwar administrations. Moreover, beginning with the National Cooperative Research Act of 1984, Congress has modified the statutory treatment of joint research and production activities. These shifts in policy were in large part a reflection of intensified competition between the largest U.S. firms and foreign enterprises. Simultaneously, the European Union has begun to adopt regional competition policy guidelines that are substantially more stringent than those of many of its member states and in a 1994 antitrust case (against Microsoft) coordinated its actions with the U.S. Justice Department in negotiating a settlement. Future U.S. antitrust policy, which aided the postwar growth of start-up firms in many high-technology sectors, may thus resemble that of other industrial regions in the future more closely than in the past.

Finally, U.S. firms have since 1980 been embarked on a far-reaching series of initiatives in restructuring their industrial research activities. For much of the postwar period, large firms such as Du Pont or General Electric relied heavily on central corporate research laboratories for the basic research that they expected to serve as the foundation for their commercial innovations. Many post-1980 corporate initiatives, however, represent efforts to "externalize" a larger fraction of their R&D activities. As a result, many central corporate research laboratories have been closed, and others have shrunk in size and as a share of corporate R&D spending. Many of the pioneers in the development of the in-house R&D laboratory now are engaged in an array of domestic and international collaborative relationships with other firms, groups of firms, government laboratories, and universities. Like the university-industry collaboration we discuss in this chapter, some of this externally oriented activity represents a revival of an important function fulfilled by many of these firms' in-house research operations before World War II – monitoring their external technological environment for innovations or inventions worthy of purchase for development and commercialization.

The implications of these changes for the future development of the U.S. R&D system are uncertain. But they are likely to reduce the structural differences between the United States' and foreign nations' R&D systems that have been so prominent for the past half century. This changing institutional landscape is an indispensable backdrop to our discussion of the technology "clusters" that have been at the center of technological and economic change within the U.S. economy during this century. As we note throughout our discussion of these clusters, a number of factors in addition to the institutional structure of the U.S. R&D system produced a unique trajectory of technological change within each of these technologies during this century.

3

The Internal Combustion Engine

THE INTERNAL COMBUSTION ENGINE, which made the automobile and the airplane possible, is often regarded as the quintessential contribution of American technology to the first half of the 20th century. Nevertheless, the initial development of the gasoline-powered engine was almost entirely a European achievement, dominated by German and French contributors – Carl Benz (a German who operated the first vehicle to be run by an internal combustion engine in 1885), Gottlieb Daimler, Nikolaus Otto, Alphonse Beau de Rochas, Peugeot, Renault, and others.[32]

The development and diffusion of the internal combustion engine illustrate a number of the broader themes that have characterized 20th-century U.S. innovation. The engine's rapid improvement and adoption within the United States were paced by the domestic abundance of low-cost petroleum-based fuels and the strong latent demand for low-cost automotive and air transportation among geographically dispersed U.S. population centers. In some contrast to the later development of new products and processes in the chemicals industry, or the post–World War II development of the electronics industry in the United States, the refinement of the internal

---

[32] An American, George Selden, applied for a patent on a gasoline engine to power an automobile in 1879 and was eventually granted such a patent in 1895. The patent provided the basis for extensive litigation, but Selden never played a role in the manufacture of automobiles.

combustion engine progressed during the early years of this century with little or no assistance from academic research.

The internal combustion engine also demonstrated the growing importance and often unexpected nature of intersectoral flows of technologies within the U.S. economy. The internal combustion engine itself was applied in a broad array of products in the transportation and other sectors (e.g., farm implements – the tractor played a pivotal role in the mechanization of cotton harvesting and the ensuing migration of farm labor from the South). The mass production methods that were perfected for the automotive industry were adapted to use in other industries. In addition, both this industry and the aircraft industry became important sources of demand for the technological advances of materials and component producers.

Automobile

The automobile – even if one overlooks its contributions to the development and diffusion of mass production – was a singular, transforming innovation. It brought with it drastic alterations in the pattern of land use. It changed the entire rhythm of urban life, including the spatial organization of work and residence, patterns of socializing, recreation, and shopping, and led to the vast expansion of suburbs. Although it was originally a European invention, the automobile was far more widely adopted, and far more rapidly adopted, in the United States than in Europe.

Indeed, the important role of the United States as the first major market for automobiles may also have contributed to the triumph of the internal combustion engine over steam and electricity, the competing sources of automotive propulsion at the dawn of this century. The emergence of internal combustion as the dominant propulsion technology was by no means a foregone conclusion in 1900, when 1,681 steam-powered automobiles, 1,575 electric cars, and

936 gasoline-fueled automobiles were manufactured in the United States (Flink 1970, p. 234). Gasoline-powered automobiles were outnumbered by steam and electric cars in the registration data for both New York and Los Angeles in early 1902. By 1905, however, the internal combustion engine was the dominant propulsion technology in the U.S. automobile industry.

All three forms of propulsion required an elaborate infrastructure for refueling or recharging, but the superior operating range of the gasoline-powered automobile gave it particular advantages in the U.S. market. In addition, the low domestic price of gasoline, relative to that of electrical power, gave internal combustion an operating-cost advantage over electric automobiles in the United States.[33] The performance of the internal combustion engine also improved more rapidly during the period from 1900 through 1905 than did electrical or steam automotive power technologies. These improvements reflected the ease with which advances in manufacturing methods (e.g., increased precision in machining cylinders, improved casting methods) could support a series of individually small, but cumulatively large, advances that enhanced the performance of the internal combustion engine. Such manufacturing advances built on the 19th-century development by U.S. firms of technologies for large-scale production of interchangeable metal parts described subsequently in this chapter.

The automobile's economic impact can be summarized by observing that the industry was classified by the U.S. Census Bureau in 1900 among the "miscellaneous" industries. The fifty-seven establishments that were entirely devoted to automobile manufacture in that year produced a total output of less than $5,000,000 in 1900 (in 1994 dollars, somewhat more than $65 million), and they were still primarily engaged in experimental work. By 1909, the automobile

---

[33] According to Flink (1970), the cost of driving an electric car from Boston to New York in 1903 was four to five times that of driving a gasoline-powered car over the same route.

Table 2. Automobile
Registrations in the United
States, 1900–93.

Year	Registrations
1900	8,000
1910	458,300
1920	8,131,522
1930	23,034,753
1940	27,165,826
1950	40,339,077
1960	61,671,390
1970	89,243,557
1980	121,600,843
1985	131,864,029
1990	143,549,627
1993	146,314,000

Sources: U.S. Dept. of Transportation 1985; U.S. Bureau of the Census (1975, 1995).

industry ranked 17th in the United States by value added, and by 1925 it ranked first. In 1900 there were 8,000 motor vehicles registered in the United States; in 1910 there were more than 458,000; and in 1930 there were more than 23 million (Clark 1929; Fishlow 1972) (Table 2). An industry that was virtually nonexistent in 1900 and ranked 17th in 1909 was ranked first (by value added) in 1925.

The 20th-century automobile, like the airplane, relied on 19th-century antecedents. These antecedents, along with the high income levels prevailing in the United States, the greater willingness of the consuming public to accept standardized products, and the immense distances that made improved transportation technologies especially attractive, had a great deal to do with the rapid growth and American domination of this new industry. These antecedents also included U.S. advances in manufacturing methods.

In the second half of the 19th century, the United States developed a large armamentarium of specialized metal-working machinery, the most important of which were machine tools. These tools catered to the production requirements of numerous industrial products, including textile machinery, railway equipment, firearms, agricultural equipment, sewing machines, and bicycles. Sequences of specialized machines produced large quantities of uniform component parts that were eventually assembled into final products whose growing technical complexity required further refinement of the "American system of manufactures." In the face of a growing demand for such products, American industry came to excel in the low-cost production of large quantities of standardized, finished products (Rosenberg 1969).

The bicycle was especially important in refining this technology and in popularizing new methods, such as the stamping of sheet steel and the use of ball bearings, that played major roles in the 20th century. This industry also made important contributions to the automobile through the development of the pneumatic tire and bicyclists' increased demand for better road surfaces. Indeed, it would be fair to say that the bicycle literally paved the way for the automobile. The development in the United States of both the airplane and the automobile were undertaken by men with extensive experience in the bicycle industry, many of whom were located in the Midwest. The growth of automobile production, however, was associated with rapid declines in the output of bicycles. Whereas 1,113,000 bicycles were manufactured in the United States in 1900, only 299,000 were produced in 1914 (U.S. Department of Commerce, 1919, vol. II, p. 753).

Henry Ford was neither the inventor of the automobile nor even the first American to experience commercial success in the sale of automobiles (this distinction belongs to Ransom Olds, who sold 500 Oldsmobiles in 1900). Nevertheless, Ford's spectacular expansion of automobile production in the second decade of the 20th century bears an interesting resemblance to later Japanese experience with transistorized radios, VCRs, or, for that matter, automobiles. Ford took an existing technology of foreign origin, redesigned it, and introduced drastic improvements in methods for its manufacture. Ford himself later described his contributions to the development of mass production methods as follows:

> As to shop detail, the key word to mass production is simplicity. Three plain principles underlie it: (a) the planned orderly and continuous progression of the commodity through the shop; (b) the delivery of work instead of leaving it to the workman's initiative to find it; (c) an analysis of operations into their constituent parts. These are distinct but not separate steps; all are involved in the first one. To plan the progress of material from the initial manufacturing operation until its emergence as a finished product involves shop planning on a large scale and the manufacture and delivery of material, tools and parts at various points along the line. To do this successfully with a progressing piece of work means a careful breaking up of the work into the sequence of its 'operations.' All three fundamentals are involved in the original act of planning a moving line production. (*Encyclopaedia Britannica*, 13th ed.).

Ford's (ghost-written) encyclopedia article succinctly captured America's central contribution to the automobile: a new production technology. This new manufacturing technology was applied to a wide range of products in the course of the 20th century, and eventually led to the eponymous term "Fordism."

Although substantial progress had been made between 1850 and 1900 in developing precision techniques that provided a higher degree of uniformity, the nascent automobile industry inherited a production technology that did not yet offer a satisfactory basis for assembling component parts of complex consumer durables at high manufacturing volumes and low unit costs (Rosenberg 1969,

1972; Hounshell 1982). Henry Ford had tentatively experimented with a conveyor-belt system for the assembly of magnetos in 1913, and these methods were quickly applied to chassis assembly. Much experimentation was needed to determine optimum assembly line speeds, positioning of workmen, the most convenient height for the performance of each task, the most efficient methods of material routing, machine layout, and so forth. Ford's seminal innovations in production organization employed a progressive assembly line, which relied on conveyor belts to move products from station to station and highly specialized machine tools that produced interchangeable parts. Ford workers assembled components that had been produced to a sufficiently high degree of precision that they required no "fitting," that is, filing, additional machining, or other operations to be inserted into the manufactured product (Raff 1991). The pace of Ford's assembly line thus was both faster and more stable than those of competitors, and the high fixed costs associated with this production organization meant that growth in the scale of production significantly reduced unit costs.

With his introduction of the Model T in 1908, Ford placed on the market a cheap four-cylinder car that was not just a new toy for the elite, as was the case in Europe, but a consumer durable for the masses. In 1912, when Ford was preparing to demonstrate to the world the possibilities of standardized, high-volume production, an influential British trade journal commented:

> It is highly to the credit of our English makers that they choose rather to maintain their reputation for high grade work than cheapen that reputation by the use of the inferior material and workmanship they would be obliged to employ to compete with American manufacturers of cheap cars.[34]

---

[34] *Autocar*, September 21, 1912, as quoted in Saul (1962, p. 41). The belief long persisted in British industry that high quality was incompatible with mass production. Much evidence on this point for the late 1920s may be found in Committee on Industry and Trade (1928, pt. 4, pp. 227–228, 220–221, and passim).

By the outbreak of the First World War America was unmistakably the home of the automobile.[35] In 1916, the Ford Motor Company sold more than half a million Model Ts at a retail price of less than $400 (in 1994 prices, roughly $5,400, less than 30% of the average price of a new U.S. automobile in 1995). Despite the apparent simplicity of Ford's production methods, they did not diffuse widely even within the automobile industry until after World War I. But the increase in U.S. automobile production to 3.6 million units in 1923 would have been utterly unattainable without the new assembly line technology. Moreover, during the 1920s these techniques were applied to a broader range of products, resulting in the rapid expansion in output of the new consumer products of the electrical industry – motors, washing machines, refrigerators, telephones, radios – as well as other consumer durables. These production methods were also applied to producer durables, such as farm machinery and equipment, and to numerous other products that could be produced in sufficiently large quantity to justify the high fixed costs that these methods required (Rosenberg 1972, pp. 106–116).

The application of these mass production methods to automobile manufacture, the single-minded pursuit of process efficiencies through more specialized capital equipment, higher levels of vertical integration, and limited modifications in the design of Ford's basic product culminated in the River Rouge complex outside of Detroit, opened in 1919 on the site of Ford's wartime shipyards. The River Rouge complex extended the concept of continuous-flow processing upstream to the raw materials for the Model T and included a

---

[35] According to Clark (1929, p. 163), "An English estimate of the number of motor vehicles in the principal countries of the world in 1914, including motorcycles which were more common in Great Britain than elsewhere, credited the United Kingdom with 426,000, France, the cradle of the automobile industry, with 91,000, Germany with 77,000, and Italy with 20,000. In the United States there were about 1,200,000 motor cars of all kinds or nearly twice as many as in all these countries combined."

large steelmaking facility (one that used iron ore from Ford-owned mines) for the production of a key input for the production of automobiles and tractors (Nevins and Hill 1957). The River Rouge site was one of the largest and most advanced examples of mass-production technology of its time, but it focused on improving efficiency in the manufacture of a single product (the Model T) whose overall design had changed little since 1908.

Ford's relentless pursuit of production efficiencies at the expense of product innovation proved to be vulnerable to a challenge from General Motors in the mid-1920s. Under the leadership of Alfred P. Sloan and former Ford production manager William Knudsen, General Motors applied many of the Ford production methods to the manufacture of common components that spanned a broader product line that could accommodate annual design changes (Raff 1991). These innovations forced Ford to the brink of bankruptcy. Ford ceased production of the Model T in 1927 and closed the River Rouge complex for most of that year in a crash effort to simultaneously develop a new product (the Model A) and retool the huge production complex. Both General Motors and another entrant, the Chrysler Corporation, benefited at the expense of Ford.

The growth of Chrysler during the 1920s was remarkable because it took place against a backdrop of higher entry barriers and increasing producer concentration. The higher capital costs of the mass-production technologies, installment purchase plans, and model changes that began to typify competition in the automobile industry during the 1920s were associated with the exit of producers, a trend that accelerated during the depression. According to Raff and Trajtenberg (1997), between 1910 and 1920 more than 150 firms were active in automobile manufacture. By the decade of the 1920s, however, this average had dropped to 90, and no more than 30 firms remained active during the period from 1930 through 1940. But the higher fixed costs and producer exit during this period were associated with a significant increase in the number of body models offered by this shrinking pool of firms – from slightly more than five body models between 1910 and 1920, the average number of models per manufacturer increased to more than eighteen by the 1930s.

The spectacular growth of the automobile industry in the first quarter of the century, as well as its immense size for the rest of the century, generated a huge demand for advanced inputs of all sorts, creating incentives for innovation in supplier industries. The automobile industry served as a kind of magnet for a diverse array of inputs: machine tools; rivers of paint for the body; immense quantities of glass, rubber, steel (including numerous alloy steels), aluminum, nickel, lead, electrical and, later, electronic components; and plastics of all sorts after the Second World War. Widespread use of the internal combustion engine in both automobile and aircraft engines sharply increased demand for petroleum products, notably fuels derived from the lighter fractions of the refining runs, with far-reaching consequences for the U.S. petroleum and chemicals industries. Virtually all of these supplier sectors experienced significant unit cost savings as more capital- and scale-intensive production methods were adopted and as incremental improvements resulted from learning in production.

In contrast to the industry's early history of rapid advances in product design,[36] the postwar U.S. automobile industry presents a portrait of a concentrated industry with little significant product

---

[36] "[T]he highest rate of quality change occurred at the very beginning [of the industry's history] (1906–14). This is undoubtedly the portion of our period in which the greatest proportion of entrepreneurs were engineers or mechanics by training, knowledge spillovers were all-pervasive, and design bureaucracies were shallowest. Whatever the mechanisms may have been, the pattern lends support to the conjecture that it is indeed in the course of the emergence of a new industry that the largest strides in product innovation are made" (Raff and Trajtenberg 1997, p. 88). Raff and Trajtenberg go on to point out that the rate of decline of quality-adjusted prices in the earliest years of the U.S. automobile industry was nearly one-half as large as those observed in the U.S. electronic computer industry during the 1980s.

innovation. The fundamental architecture of the automobile was achieved by roughly 1925 – an enclosed steel body mounted on a chassis, powered by an internal combustion engine. And by the end of the 1930s, as Raff and Trajtenberg show in their analysis of change in the performance and other attributes of automobiles, improvement in product characteristics had virtually ceased. Writing in 1971 about innovation in the postwar U.S. automobile industry, one scholar argued that

> The auto industry can be described as a technologically stagnant industry in terms of its product. Cars are not fundamentally different from what they were in 1946; very little new technology has been instigated by the industry. The product has improved over the last twenty years, but these have been small improvements with no fundamental changes. The sources for these improvements have often been the components suppliers, rather than the auto companies themselves; and the auto companies have been slow to adopt these improvements. (White 1971, p. 258).

Although it was not innovative with respect to product designs, however, the U.S. automobile industry did continue to improve production technologies, reflected in its above-average labor productivity growth during this period.

This situation of limited product variety and innovation began to change during the 1970s, as a result of the sharp increases in the price of gasoline and the related rapid growth of foreign imports, largely from Japan. The origins and results of this competitive crisis for U.S. automobile firms are noteworthy for what they suggest about the continued importance of international transfers of "hard" and "soft" technologies among industrial economies. By the late 1970s, leading Japanese automobile firms such as Toyota and Honda had perfected new techniques for production organization and product development (some of which, especially in the area of "quality management," relied on statistical and management techniques originally developed by U.S. managers and promoted within Japanese industry in the aftermath of World War II) that made possible the creation and manufacture of a broader variety of higher-quality products than were available from U.S. producers (Clark and Fujimoto 1991; Womack, Jones, and Roos 1990). The resulting dislocations within the U.S. automobile industry resulted in the imposition of restrictions on Japanese automobile imports by the U.S. government, which in turn produced a wave of Japanese investment in new automobile production facilities in the United States. These "transplant" factories eventually served as important channels for international technology transfer.

The success of Japanese firms in applying so-called "lean production" techniques in their U.S. automobile plants provided compelling and credible evidence to U.S. managers that alternative approaches to production organization were both feasible and profitable with a U.S. workforce. Faced with growing demands from consumers for greater fuel economy and safety, as well as government-imposed requirements for reductions in pollution, automakers designing products for the U.S. market were also compelled to redesign engines and transmissions, significantly increasing their use of semiconductors and electronic components, including integrated circuits, microprocessors, and computers, in automobiles. In 1996, the North American automotive industry consumed nearly $3 billion in semiconductor components, and this sector's consumption of integrated circuits alone (nearly $2 billion) outstripped that of the U.S. defense industry.[37]

Although the "dominant design" in the automobile industry remains remarkably similar to its antecedents in the 1920s and 1930s, significant innovations have occurred since 1975 in components that rely heavily on the "import" of technologies from other industries (among other things, the more extensive application of electronics to automobiles has raised the skill and training requirements

---

[37] We are grateful to Jeffrey Macher of the Haas School of Business for collecting these data and to Tier One, Inc. for granting permission to use them.

for mechanics significantly – see Stern [1997]). The postwar history of the U.S. automobile industry thus illustrates the importance of these intersectoral technology flows, as well as the importance of international flows of products and capital in transferring even soft technologies for the organization of production and product development. The postwar history of product innovation in the U.S. automobile industry also suggests the importance of competitive pressure in maintaining innovative performance. Although in many respects this evidence of the importance of competition among even large, oligopolistic firms is consistent with the arguments of *Capitalism, Socialism and Democracy* (Schumpeter 1942), the evidence from 1945 through 1975 suggests that domestic oligopolies may succumb to the pursuit of the quiet life, rather than maintaining their investments in creative destruction.

The Airplane

The internal combustion engine gave birth not only to the automobile but also to trucks, buses, and other commercial vehicles, agricultural equipment, and the airplane.[38] The invention of the airplane is indelibly associated in the public mind with the brief flight of the Wright brothers' clumsy contraption at Kitty Hawk in 1903. The lifting off the ground of a fixed-wing, heavier-than-air, powered device for even a few seconds was indeed an extraordinary achievement in the history of technology. But like most major inventions, this flight was a harbinger of things to come, rather than a demonstration of a device of any immediate utility. Although the airplane played a significant role in World War I, the biplane design that dominated wartime production was not suitable for commercial transport purposes. Considerable time elapsed after the

---

[38] Internal combustion engines were also important on ships and railroads, but they were eventually displaced in these uses by diesel engines.

breakthrough at Kitty Hawk before the airplane became an invention of genuine economic significance, just as Benz's demonstration of the automobile engine preceded the automobile industry's emergence as an important economic sector by roughly two decades.[39]

The end of World War I brought with it a precipitous decline in the output of military aircraft. Whereas more than 14,000 aircraft were produced in 1918, total U.S. production amounted to less than 300 aircraft in 1922 (Holley 1964). Production began to revive with the military's announcement of plans in 1926 to expand its aircraft fleet to 26,000 planes by 1931. The Kelly Air Mail Act of 1925 transferred responsibility for transportation of airmail from the U.S. Post Office to private contractors, and federal airmail contracts incorporated subsidies for the adoption of new commercial aircraft technologies, such as multiengine aircraft, radio, and navigational aids. The National Advisory Committee on Aeronautics, formed in 1915, sponsored important research on airframe design, and the founding of Pratt and Whitney in 1925 was based on the U.S. Navy's interest in purchasing its Wasp engine. Construction of the infrastructure for a civilian transportation system, including radio networks and aerial beacons, also began during the 1920s.

By the late 1920s, the growth in American passenger demand offered prospects of a large commercial market, and between 1927 and 1937 this market accounted for 42 percent of U.S. aircraft sales (Miller and Sawers 1968, p. 2). In fact, American passenger traffic, which was almost nonexistent in 1927, was larger than that of the whole of the rest of the world in 1930 (Miller and Sawers 1968, p. 16). The rapid expansion of this mode of commercial transportation, like the adoption of the automobile, reflected the

---

[39] This observation underlines a fundamental aspect of the life history of many technologies, especially those discussed here: Most are of limited use at birth and need to undergo extensive performance improvement, design modification, and cost reduction before they can exercise a significant impact on the economy (Rosenberg 1994).

long distances associated with domestic travel in the United States, as well as a large and affluent population.

The revival of the aircraft industry during the 1920s was associated with a series of mergers that produced for the first and only time in the history of the industry several vertically integrated firms that combined air transportation, airframe production, and engine manufacture.[40] The Air Mail Act of 1934, passed by Congress in reaction to political controversies over the Post Office's awarding of contracts, mandated the dissolution of these integrated aircraft and air transportation firms, as well as the termination of subsidies for adoption of advanced aircraft technologies. Demand for commercial air travel both contributed to and was enhanced by the introduction in 1936 of the DC-3, easily the most popular commercial aircraft ever built. The DC-3, a twenty-one–passenger aircraft, carried 95% of all commercial traffic in the United States by 1938 and was used by thirty foreign airlines. Including the numerous military variants of this airplane built during the Second World War, the total number of DC-3s produced exceeded 13,000 (Miller and Sawers 1968, p. 103).

No single, "critical" technical improvement accounted for the astonishing commercial success of the DC-3. Like the Model T, the commercially dominant DC-3 was a synthesis of technological advances in a diverse array of components and materials technologies that underwent steady modification and improvement in design and manufacturing long after its commercial introduction. The aircraft incorporated a large number of specific inventions and design improvements, originally developed both in the United States

---

[40] United Aircraft, founded in 1929, was formed from a merger of Boeing Aircraft, Boeing Air Transport, Pratt and Whitney, Chance Vought Aircraft, the Hamilton Standard Propeller Corporation, and Stearman Aircraft. North American Aviation, founded in 1928, included Curtiss Aeroplane and Wright Aeronautical and controlled large minority shareholdings in Transcontinental Air Transport and Western Air Express (two airlines that subsequently merged to form TWA).

and Europe, many of which it shared with other aircraft, including Douglas's own earlier DC-l and DC-2. Yet the DC-3 brought together many interdependent and mutually reinforcing features: a two-engine aircraft incorporating numerous improvements in engine design (e.g., air-cooled radial engines with cowlings to reduce drag) that relied on fuels that would permit higher compression ratios, variable-pitch propellers, wing flaps, streamlined monoplane design, retractable landing gear, cantilevered wings, and stressed-skin (monocoque) multicellular metal construction.

The elements of an optimal aircraft design are extremely complex, because aircraft design is inherently full of trade-offs that need to be mediated, and the appropriate criteria for such trade-offs are often uncertain in the development of such products. The performance of any single component depends heavily on other components, and the exact nature of that relationship historically has not been predictable by recourse to scientific theory. Thus, although the airplane was already a third of a century old in 1936, it would be fair to say that the DC-3 represented the most important innovation in the history of commercial aircraft up to that time. One indicator of the advance in economic performance represented by the DC-3 is its cost per available seat mile – the DC-3 design represented a decisive cost improvement over its immediate predecessors (Table 3). Moreover, it was also superior in terms of comfort, safety, and reliability.

The success enjoyed by U.S. commercial aircraft firms during the interwar period is remarkable, because until the Second World War, most of the leading scientific research in aerodynamics was performed in Germany, rather than the United States. Theoretical advances in this realm were dominated by the research of Ludwig Prandtl of the University of Gottingen. During his lengthy research career (1904–53), Prandtl provided the analytical framework for understanding the fluid mechanics that underlie the flight performance of aircraft. U.S. aeronautical engineering research at the California Institute of Technology, Stanford, and MIT drew

Table 3. Comparative Operating Costs of Leading Interwar Commercial Transports.

| | | | Comparative Operating Costs, Cents per Available Seat-mile | | | | | |
| | Date of Introduction | No. of Passenger Seats | Flying Operations | | | Direct Maintenance | Depreciation | Total |
Aircraft			Flight Personnel	Fuel and Oil	Total*			
Ford Trimotor	1928	13	0.72	0.47	1.34	0.67	0.62	2.63
Lockheed Vega	1929	6	1.01	0.28	1.56	0.58	0.37	2.51
Boeing 247	1933	10	0.74	0.36	1.19	0.43	0.49	2.11
Douglas DC-3	1936	21	0.34	0.28	0.69	0.24	0.34	1.27

Source: Miller and Sawers (1968), p. 34.

* The "Total" figure for flying operations includes other miscellaneous costs (e.g., insurance and damage reserve provisions) in addition to personnel, fuel and oil.

heavily on Prandtl's fundamental work. Indeed, aerodynamics may be said to have come to America in the person of Theodore von Karman, Prandtl's most distinguished student, who emigrated to the United States in the late 1920s to take a research position at Cal Tech (Hanle 1982).

Despite their limited role in theoretical aerodynamics research, U.S. universities were important to America's rise to technological leadership in aircraft. The growing role of U.S. universities as sites for experimentation and design research during the pre-1940 period reflected their improved research capabilities. Much of this university work consisted of extensive testing, relying on experimental parameter variation because no scientific theory provided detailed guidance on the design of aircraft. Nevertheless, as Vincenti has suggested, experiments such as those on propeller design by Durand and Lesley at Stanford during the period from 1916 through 1926 represented more than just data collection, albeit something other than science. Their experiments relied on a specialized methodology that could not be directly deduced from established scientific principles, although it was obviously not inconsistent with those principles. These experiments were not fundamental science, but neither can they be accurately characterized as applied science:

> [T]o say that work like that of Durand and Lesley goes beyond empirical data gathering does not mean that it should be subsumed under applied science... [I]t includes elements peculiarly important in engineering, and it produces knowledge of a peculiarly engineering character and intent. Some of the elements of the methodology appear in scientific activity, but the methodology as a whole does not. (Vincenti 1990, p. 166).

This work formed the basis for important, incremental advances in engine design and performance. The research is a good example of the use of applied engineering research to analyze and describe an important phenomenon in the absence of a comprehensive scientific theory.[41]

The next major innovation in aircraft after the DC-3 was the jet engine, application of which to commercial transports over the course of the 1950s and 1960s ended Douglas Aircraft's dominance of the commercial aircraft industry. Early work on the jet engine was performed in Britain and Germany during the 1930s, in anticipation of its use in military aircraft. Consistent with our characterization of the United States as lagging behind the scientific frontier in many areas before 1940, U.S. weakness in theoretical aerodynamics meant that prewar industry, government, and academic researchers in this country were slow to recognize the potential and feasibility of jet-powered aircraft. Jet engine technology reached the United States from Great Britain during the war, when General Electric developed military engines – as in other episodes of "technology transfer," a great deal of time and energy were required to convert the vast quantity of British blueprints and technical diagrams to American specifications.[42] Codified knowledge alone was insufficient to enable U.S. producers to duplicate the British innovation.

Although it was used in military applications in the closing months of World War II and in the Korean War, the jet engine did not enjoy success in civilian markets until the late 1950s. Illustrating the risks of being "first to market" in applying a new technology that has significant uncertainties in its performance, the introduction of commercial jet service by British Overseas Airline Corporation in 1952 with the DeHavilland Comet was a disaster. The failure to predict metal fatigue in the DeHavilland design resulted in a series of crashes and the commercial failure of the

---

[41] This type of work is by no means confined to pre–World War II U.S. R&D, as we note in our discussion of the postwar electronics revolution.

[42] "When the British in World War II supplied us with the plans for the jet engine, it took ten months to redraw them to conform to American usage" (Arrow 1976, p. 174).

aircraft. Boeing launched its first commercial jet, the 707, in 1958, followed shortly by the Douglas DC-8.

Pratt and Whitney, a leading producer of piston aircraft engines, entered the production of jet engines shortly after World War II. In spite of development efforts by Westinghouse and the Allison division of General Motors, General Electric and Pratt and Whitney dominated the commercial jet engine market by the 1960s, by which time the entire commercial engine market (excluding private aircraft) had become a jet-engine market. The commercial application of jet engines was associated with the exit of U.S. producers of aircraft engines and airframes, as well as the loss of their interwar commercial dominance by Douglas and Lockheed. By the 1980s, only two U.S. producers of airframes (Boeing and McDonnell Douglas) and two U.S. producers of engines (Pratt and Whitney and General Electric) remained, and in 1997, Boeing merged with McDonnell Douglas. The appearance of the radically new technology of jet engines had transformed the structure of the commercial aircraft industry. In contrast to the transformation of other postwar high-technology U.S. industries, however, commercial aircraft innovators in the jet age were not new firms and in most cases were not new entrants to the commercial aircraft industry (General Electric is one exception).

During much of the postwar period, the U.S. commercial aircraft industry benefited from its close links with an important defense industry, military aircraft and engines. U.S. military expenditures on R&D accounted for more than 70% of industry R&D spending during the postwar period, although the composition of those expenditures changed drastically with the development of missile technology. Technological spillovers from military to civilian applications of jet engines, materials, and electronics aided the U.S. commercial aircraft industry. The industry also benefited from military procurement of airframes and engines – a portion of the costs of developing the Boeing 707 were borne by the earlier development of the KC-135, a jet-powered military tanker. In the late 1960s, perhaps the single most important of all postwar improvements in jet engines came from military sources. R&D supported by the Pentagon on jet engines for the giant C-5A transport led to the development of the high–bypass ratio engines that now power many commercial transports. Since the 1970s, however, the economic significance of these military-civilian spillovers has declined in the commercial aircraft industry. Indeed, the most recent military tanker, the KC-10, was based on a civilian airframe design, the McDonnell Douglas DC-10.

Like the automobile industry, the U.S. commercial aircraft industry has benefited throughout its history from technological innovation in other industries. The monocoque airframe of the DC-3 used duralumin, developed by Alcoa in Navy dirigible programs back in the 1920s. The advent of the jet engine meant that metallurgy assumed substantial importance for advances in propulsion technology. Since the 1940s, research on the behavior of metals at high temperatures has contributed to the development of turbine blades, inlets, outlets, and compressors for turboprop and jet engines. General Electric, a major producer of steam turbines and other power generation equipment, became involved in metallurgical research for the development of supercharged aircraft engines and, later, jet engines. Improvements in the performance of these propulsion technologies, as well as in piston aircraft engines, also relied on advances in fuels resulting from R&D sponsored by automotive and petroleum firms.

Both U.S. airlines and the U.S. commercial aircraft industry have benefited from postwar advances in electronics. Growing use of semiconductors was spurred by the requirements of strategic missile guidance systems in the 1950s. The postwar threat to the continental United States posed by strategic bombers provided a significant impetus to Cold War programs of support for the computers and electronic systems that utilized semiconductors and software. Compared with vacuum tubes, solid-state circuits were far lighter and more reliable and generated less heat. The increased importance of military space projects, many of which were carried out by aircraft firms, blurred the boundaries between the electronics and aircraft industries. Although semiconductor-based guidance systems produced substantial benefits for commercial aircraft, the origins of semiconductors themselves were remote from the commercial aircraft industry, stemming from Bell Telephone Laboratories' efforts to improve long-distance telephony.

The computer has been a source of numerous improvements throughout air transportation and the commercial aircraft industry. It is essential to air-traffic control and to the determination of optimal flight paths that, aided by information from weather satellites, have saved energy and improved passenger safety and comfort. Computers have made possible the worldwide ticketing and reservation systems that are at the heart of large airlines' pricing and scheduling strategies. Cockpit minicomputers have significantly improved the navigation and maneuvering performance of commercial aircraft, and computer simulation is now the preferred method for teaching neophytes how to fly.

Aircraft and engine design and development have also been transformed by the widespread use of powerful computers. Computer-assisted design techniques have reduced, although they have not eliminated, uncertainties over airframe performance, enabling more extensive testing to be carried on outside of wind tunnels. Supercomputers have played an important role in the wing designs of most recent commercial transports. Indeed, the Boeing 777, which entered commercial service in 1996, was designed largely by computer-aided techniques that linked Boeing designers with both prospective customers and suppliers; computers also were used to an unprecedented extent in testing the design.

The contributions of the computer to air transportation raises a much more general point concerning 20th-century technological change. The research that is responsible for technological improvements tends to be highly concentrated in a small number of industries, but each of these few industries generates technologies that often are widely diffused throughout the economy. Thus, R&D within an industry is a necessary but by no means sufficient condition for technological change in that industry. Although the aircraft industry's "own" R&D has been essential to the absorption and application of technologies developed outside its boundaries, research outside the industry has been at least as important a source of performance improvement as research carried out within it.

The application of computer-aided design and simulation technologies, for all their labor-saving potential, does not appear to have significantly lowered the costs of developing new airframes and engines. Indeed, one of the hallmarks of the postwar commercial aircraft industry in the U.S. and other industrial economies has been the inexorable increase in the costs of developing new products, which now exceed $3 billion for a new large commercial transport airframe and a similar amount for a new commercial jet engine. Moreover, these increases in development costs have occurred against a backdrop of declining military-civilian technological spillovers, increasing the share of these development costs that must be borne by private firms.

The increased costs of new product development have been associated with exit from the U.S. commercial aircraft industry since 1945 and at least one federal government–guaranteed "bailout," of Lockheed Corporation in 1975. The remaining firms have reorganized their new product development activities and now rely on a complex array of "alliances," multifirm joint ventures with non-U.S. firms, for the development of new airframes and engines. Such alliances are necessary to share risk and also facilitate the penetration of the rapidly growing foreign markets for airframes and engines. The scope of this U.S. industry's markets and of its innovative activities now are global, and significant two-way transfers of technology occur within these interfirm alliances (Mowery 1987).

Conclusion

This chapter has discussed only a portion of the applications of the internal combustion engine within the 20th-century U.S. economy. Its progressively widening areas of application meant that the internal combustion engine led to upheaval and laid the basis for new industries in sector after sector. In addition to automobiles and commercial aircraft, this technology transformed agriculture, through its application to tractors and harvesters. The application of internal combustion engines to commercial vehicles, such as trucks, contributed to the growth of new approaches to the distribution of foodstuffs and other consumer goods that, along with other innovations, ultimately contributed to a considerable reorganization of the retailing industry. In most U.S. manufacturing industries, however, the adoption of the internal combustion engine eventually produced a relatively stable producer structure and pattern of incremental technological change.*

---

* In the commercial aircraft industry, as we note in this chapter, it was the replacement of the piston by the jet engine that transformed the industry's structure.

4

Chemicals[43]

THE U.S. CHEMICALS INDUSTRY, like the aircraft and automobile
industries, has benefited throughout this century from scientific and technological advances originating elsewhere in the global
economy. The primary contributors to fundamental knowledge of
chemistry in the early decades of the century were virtually without exception Europeans. In the course of the century, however,
the American scientific contribution grew, and since 1945 (in no
small measure as a result of events connected with that war), the
center of fundamental chemical research has been located in the
United States. A comparison of trends in awards of the Nobel Prize
in Chemistry to citizens of the United States and the major European powers before and after 1940 is revealing in this connection. Through 1939, German scientists received fifteen out of the
thirty Nobel Prizes awarded in chemistry, U.S. scientists received
only three, and French and British scientists each accounted for
six. Between 1940 and 1994, U.S. scientists received thirty-six of
the sixty-five chemistry Prizes awarded, German scientists received
eleven, British scientists received seventeen, and French scientists
received one (*Encyclopaedia Britannica*, 15th ed., pp. 740–747).

A central feature of technological change in chemicals during this
century was undoubtedly the rise of the petrochemical industry, that

---

[43] Portions of this chapter draw on Rosenberg (1998a) and Arora and Rosenberg
(1998).

is, the shift in organic chemicals away from a feedstock based on coal to one based on petroleum and natural gas. American leadership here was overwhelming,[44] and once again, the U.S. natural resource base played an important role in guiding the development of petroleum-based chemicals processes by domestic firms. But German scientific and technological capabilities also shaped American technological developments.

The German chemicals industry throughout the 1890–1945 period focused on the development of synthetic products. German capabilities in synthetics owed much to the scientific and technological sophistication that had been generated before World War I in the synthetic-dye industry.[45] Both the German resource endowment and domestic concerns over dependence on foreign sources of feedstocks dictated that these synthetic products were derived from coal rather than petroleum.

Technological change in the American chemical industry has been shaped by several features: the large size and rapid growth of the American market, the opportunities afforded by large market size for exploiting the benefits to be derived from large-scale, continuous-process production, and a natural resource endowment – oil and gas – that provided unique opportunities for transforming the resource base of the organic chemical industry and achieving significant cost savings, if an appropriate process technology could be developed. During the pre-1945 period, technological change in both Germany and the United States was responsive to sharp differences in domestic natural resource endowments in an era during which political developments militated against extensive reliance on foreign supplies of feedstocks. The United States

---

[44] Williams (1982, p. 138) states simply that "the petrochemical industry was essentially an American development."

[45] On the eve of the First World War, eight German firms, along with their foreign subsidiaries, "produced 140,000 tons a year of dyestuffs out of a world total of 160,000 tons a year; and 80 percent of this tonnage was sold abroad" (Aftalion 1991, p. 104).

created new technologies that intensely exploited its abundance of liquid feedstocks, and the German chemicals industry fashioned new technologies that compensated for their absence. During World War II, German tanks and airplanes were fueled by synthetic gasoline and ran on tires made from synthetic rubber derived from coal feedstocks. Only in the wake of World War II and the creation of a set of multilateral institutions governing international trade and finance did the revival of international trade and U.S. guarantees of access to foreign petroleum sources support a shift by the German chemicals industry to petroleum feedstocks (Stokes 1994).

Synthetic Ammonia: German Leadership and "Technology Transfer"

By the early 20th century, the United States had a large chemical industry that concentrated on the production of inorganic chemicals and explosives. Measured by one widely used yardstick, the output of sulfuric acid, the U.S. industry in 1914 was almost as large as those of Germany and Great Britain combined (Haynes 1945).[46]

R&D in the U.S. chemicals industry was modest in scale and scope during this period. U.S. producers of organic chemicals still depended on natural inputs and did not remotely compare in their technical sophistication with German firms, on whom the United States depended for dyestuffs. In fact, at the outbreak of war, there were only two significant domestic producers of dyes for the huge U.S. textile industry, and their meager 3,000 tons of annual output accounted for no more than one eighth of the nation's peacetime requirements. The wartime termination of German synthetic dye imports, along with a parallel increase in domestic demand, were

---

[46] For a detailed description of the American chemical industry at the outbreak of the First World War, see Aftalion (1991, pp. 115–119).

decisive events in the emergence of an American organic chemical industry (Aftalion 1991, pp. 115–119). Indeed, the cut-off of German chemicals imports to the United States led to a crash program, sponsored by the federal government, to develop alternative sources of supply for nitrogen and ammonia.

Through much of the period from 1900 through 1940, German chemicals firms pursued the development of high-pressure technologies for the production of synthetic ammonia and gasoline, in response to constraints imposed by natural resource endowments and political concerns.[47] German technical leadership in the chemicals industry during this period was underscored by one of the most important technological breakthroughs of the 20th century: the Haber-Bosch process for nitrogen fixation, developed and commercialized by the Badische Aniline und Soda Fabrik (BASF). Development of this process in 1913 relieved Germany of her dependence on Chilean nitrates, which had been critical both to agriculture and the military. The ammonia-synthesis technology required great skills in high-pressure technologies, skills that were subsequently employed in Germany to convert abundant coal and lignite into such products as synthetic gasoline (Aftalion 1991, pp. 133–134).

The Haber-Bosch process involved the use of high pressure (up to 1,000 atmospheres) and high temperatures (up to 500°C), as well as the employment of a catalyst to produce ammonia. The catalyst was critical, and BASF maintained strict and effective secrecy on its composition and use in all of its patent filings.[48]

---

[47] This development effectively illustrates the path-dependent nature of much technical change that is also apparent in the U.S. chemicals industry during this century.

[48] "The Badische Company had effectively bulwarked this discovery with strong, broad patents which detailed meticulously the apparatus, temperatures, and pressures, but cleverly avoided particulars as to the catalysts employed or their preparation. This last information was the core of the process so far as its practical operation was concerned" (Haynes 1945, pp. 86–87).

The attempt to introduce the Haber-Bosch process for nitrogen fixation into the United States during World War I provides a classic account of the difficulties involved in international technology transfer, even into a recipient country with considerable technological capabilities. The failure of this "transfer" meant that the availability of nitrogen was a critical problem to the Allied powers, including the United States, throughout the war.

Despite the governmental expropriation of the U.S. patents of BASF and other German chemicals firms by the Alien Property Custodian in 1918, after the United States had entered the war, U.S. experts could not replicate the Haber-Bosch process for nitrogen fixation. A wartime program at Muscle Shoals, Alabama, that consumed more than $70 million (more than $500 million in 1997 dollars) proved insufficient to develop a successful process for nitrogen fixation and the production of synthetic ammonia until 1921.[49] A great deal of additional research by the U.S. Fixed Nitrogen Laboratory at Muscle Shoals and by private industry was needed during the 1920s to provide the necessary design and construction information needed for the high-pressure equipment (such as large compressors) that was essential to the widespread use of the Haber-Bosch process. Equally important was the mastery of catalytic technology, which eventually proved to be the key to the growth of the chemical industry. A prolonged learning experience was necessary to understand the two sides of catalysis, the chemical side and the engineering and design side, especially the complex process of bringing catalytic techniques from the laboratory stage to the very different circumstances of commercial-scale production (American Chemical Society 1973, pp. 216–220; Haber 1971, pp. 205–206).

---

[49] See Haynes (1945) and Hughes (1983) for accounts of the costs and difficulties encountered by the U.S. program.

Table 4. *Commercial Fertilizers: Quantities and Varieties Consumed, 1940–1985.*

Year	Quantity, 1000s of Tons	Nitrogen, %	Phosphoric Oxide, %	Potash, %
1940	8,556	4.9	10.7	5.1
1950	20,345	6.1	10.4	6.8
1955	21,404	9.0	10.5	8.8
1960	24,374	12.4	10.9	8.9
1965	33,071	16.1	11.8	9.7
1970	39,902	20.4	12.0	10.6
1975	40,630	21.2	11.1	11.0
1980	50,491	22.6	10.8	12.4
1985	47,179	24.4	9.8	11.7

Sources: U.S. Bureau of the Census, *Statistical Abstract of the United States*, various years (Washington, DC: U.S. Government Printing Office); U.S. Department of Agriculture, Crop Reporting Board, *Commercial Fertilizers* (Washington, DC: U.S. Government Printing Office, 1985).

Only after the Second World War did numerous additional process improvements, many of which relied on cheap, abundant electric power, make synthetic ammonium nitrate the leading source of fertilizer nitrogen. Eventually, its ease of shipment, distribution, and application meant that ammonia itself was directly injected into the soil in the form of anhydrous ammonia, aqua ammonia, or nitrogen solution. The great post-1945 growth in agricultural productivity in the United States and, eventually, throughout the world owed an immense debt to the increased use of chemical inputs, including not only synthetic nitrogenous fertilizers but also herbicides and insecticides (Achilladelis, Schwarzkopf, and Cines 1987).

These improvements in process technologies reduced the price of fertilizer by comparison with the prices of other agricultural inputs or prices for agricultural output, spurring the use of fertilizer.[50] The quantities of commercial fertilizer inputs into American agriculture grew more than four-fold between 1940 and the mid-1960s (Table 4). But the falling price of fertilizer had another influence on increases in output per acre. Plant breeders during the 20th century have developed a number of new and more productive seed varieties, including such important advances as hybrid corn, which began to sweep through the Midwest in the late 1930s. These new plant strains were highly responsive to fertilizer inputs, and both the development and adoption of high-yielding crop varieties were closely connected to the declining cost of fertilizers (Hayami and Ruttan 1971, p. 121). By dramatically shifting the relative prices of key inputs, innovation in chemicals created incentives for the pursuit of a particular trajectory of technology development and adoption in a very different sector of the economy.

The Development of a Petroleum-Based Chemicals Industry in the United States

The introduction and rapid adoption of the internal-combustion automobile in the opening years of the 20th century brought in its wake an almost insatiable demand for liquid fuels. This demand in turn spurred the growth of a new sector of the petroleum-refining industry in the first two decades of the 20th century that was specifically calibrated to accommodate the needs of the automobile. Petroleum refining had two important, related features. First, it was highly capital-intensive; by the 1930s it had become the most capital-intensive of all American manufacturing industries.

---

[50] See Griliches (1958) and Sahota (1968). The motivation to raise output per acre through more intensive applications of fertilizer was also, of course, strengthened by government programs involving acreage restrictions.

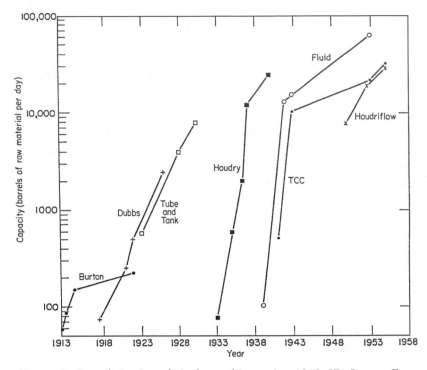

Figure 2. Growth in size of single cracking units, 1913–57. *Source:* Enos (1962).

Second, productive efficiency required that small-batch production, so characteristic of other chemical products such as synthetic organic materials, be discarded in favor of large-volume production methods. Large-scale petroleum refining required the development of continuous-process technologies. American leadership in petroleum refining provided the critical knowledge and the engineering and design skills to support the chemical industry's shift from coal to petroleum feedstocks in the interwar years.

The large size of the American market had introduced American firms at an early stage to the problems involved in the large-volume production of basic products, such as chlorine, caustic soda,

soda ash, sulfuric acid, and superphosphates. This ability to deal with a large volume of output, and eventually to do so with continuous-process technology, was to become a critical feature of the chemical industry in the 20th century.[51] In this respect, the early American experience with large-scale production contributed to the U.S. chemical industry's transition to petroleum-based feedstocks.

The development of large-scale production facilities in the United States, however, reflected more than just the incentives created by a large and growing market. American firms' expertise in the construction and operation of large-scale chemicals plants was based as much on careful empiricism as on scientific expertise. New technologies were first tested on a small scale, commonly in a pilot plant. As more reliable design data were generated, and as confidence in the new technology grew, chemical firms expanded the scale of their production facilities (Fig. 2).[52]

The dominant participants in this industrial transformation were Union Carbide, Standard Oil of New Jersey, Shell, and Dow. But the shift to petroleum depended as well on the adoption by U.S. petroleum firms, notably Humble Oil (an affiliate of Standard Oil of New Jersey), of new exploration techniques. Beginning in the mid-1920s, Humble began to use geophysical techniques for exploration that had been developed and first applied in the United States by European geologists. The results were remarkable – from

---

[51] One authoritative study, discussing the American situation shortly before its entry into the First World War, referred tellingly to "... the American attitude to the size of chemical works, which was, in short, to build a large plant and then find a market for the products." (Haber 1971, p. 176.)

[52] The same practice can be observed with respect to new generations of commercial aircraft. The "stretching" of fuselages to accommodate a larger number of passengers is a common phenomenon, but only after an interval of time long enough to establish a high degree of confidence in the design and especially in the engines.

1920 through 1926, seventy major oil fields were discovered.[53] The American exploitation of technologies developed abroad thus expanded the economically relevant resource endowment of the economy, and the development of an array of chemical technologies further increased the value of U.S. oil and natural gas deposits. Once the new processing technologies had been developed, the growing availability of low-cost petroleum and natural gas meant that these sources could provide organic chemicals far more cheaply than coal:

> Between 1921 and 1939, the production of organic chemicals *not derived from coal tar* [author's emphasis] rose from 21 million pounds valued at $9.3 million to 3 billion pounds, with a value of $394 million. Coal tar chemicals production in 1939 ... still amounted to only 303 million pounds, valued at $260 million. The average 1939 price for petrochemicals was, in fact, 13 cents a pound versus 87 cents for coal tar derived chemicals. The average price of noncoal tar chemicals had, over the period, been reduced by a factor of three, from a 1921 level of 43 cents per pound. (Spitz 1988, pp. 67–68).

Seen from the perspective of the 1990s, the intimate linkage between the petroleum and chemical industries seems natural and

---

[53] "The significance of the use of geology and geophysics is shown by historical statistics on the discovery of new oil fields in the United States. In sixty years before 1920, sixty-eight major fields had been discovered. 'Practical men,' as the old-fashioned unscientific prospectors were called, had made most of the discoveries for several decades, but geologists had gradually risen to considerable importance. The two groups were probably about equally responsible for discoveries made during World War I. In the years 1920 through 1926, geologists had been more productive than practical men; they had found two-thirds of seventy major fields. From 1927 through 1939, of 171 major discoveries, geophysicists found 65, geologists 77, and the old type of prospector found 29. It is significant also that practical men had only one successful strike out of seventeen wells drilled, as compared to the technologists' one in every 7.5" (Larson, Knowlton, and Popple 1971, p. 75).

therefore inevitable. But from the vantage point of 1920, this linkage was anything but obvious – in fact, the "natural" connection between the two industries was a complex human creation. In 1920 petroleum was regarded as a fuel and a lubricant; the chemical industry thought of its inputs in terms of chemicals in a less-processed state on the one hand, and feedstocks drawn from coke-oven byproducts on the other. Only gradually did U.S. oil companies begin to realize that their refining operations could produce not just fuel and lubricants but organic chemical intermediates as well (Spitz 1988, ch. 2).

The transformation of the U.S. chemical industry during the period from 1920 through 1946, which laid the foundation for the petrochemical industry that matured in the post–World War II years, was in large measure the achievement of the chemical engineering profession.[54] Because of the importance of chemical engineering for the transformation of the U.S. petroleum and chemicals industries, and because the development of this academic discipline illustrates the evolving relationship between U.S. universities and industry during the pre-1940 era, it merits closer attention.

Chemical engineering sought to fuse an understanding of chemistry with the process technologies necessary to produce petroleum-based products in unprecedented volume. This approach contrasted with technical practices in the German chemicals industry, which maintained sharp distinctions between chemists and mechanical engineers, the latter group being charged with development of process technologies. Significantly, the German approach to organizing process and product innovation had developed during the dyestuffs era, one characterized by small-volume, batch-production methods.

---

[54] See Landau and Rosenberg (1992). The following paragraphs draw on this account.

Integration of process and product technologies was less critical in this environment.[55]

The development of chemical engineering was associated largely with a single U.S. university, MIT.[56] Teaching and research in chemical engineering at the Institute began during the 1888–1915 period and involved considerable controversy over the nature of MIT's relationship with the evolving U.S. chemicals industry. Arthur Noyes, an MIT graduate and holder of a PhD from Leipzig University, established a Research Laboratory of Physical Chemistry in 1903 to support fundamental scientific research in chemistry. Noyes's approach to academic research was ultimately defeated, and Noyes departed for the California Institute of Technology, after Professor William Walker and Arthur D. Little, founder of the well-known consulting firm, developed a curriculum for engineering training at MIT that emphasized applied science and close links with industry. Walker founded the Research Laboratory of Applied Chemistry in 1908, in order to obtain research contracts from industry and thereby to provide both income and experience in industrial problem solving for faculty and students.

---

[55] Warren Lewis, one of the founders of U.S. chemical engineering, characterized the German situation as follows: "Details of equipment construction were left to mechanical engineers, but these designers were implementing the ideas of the chemists, with little or no understanding of their own of the underlying reasons for how things were done. The result was a divorce of chemical and engineering personnel, not only in German technical industry but also in the universities and engineering schools that supplied that industry with professionally trained men" (Lewis 1953, pp. 697–698).

[56] Although MIT was the most important academic contributor to the development of chemical engineering, a number of other universities – the University of Illinois, the University of Minnesota, the University of Wisconsin, the University of Delaware, and others – made significant contributions. Industrial firms, especially Du Pont, also played a key role. For an illuminating analysis of Du Pont's contribution, see Hounshell and Smith (1988, ch. 14).

Links between MIT and the U.S. chemicals industry were further strengthened by the foundation in 1916 by Walker, along with Arthur D. Little and Warren Lewis (a colleague of Walker's), of the School of Chemical Engineering Practice. The School emphasized cooperative education in chemical engineering, in which students spent a portion of their undergraduate years in chemicals industry firms.

The discipline of chemical engineering that was developed at MIT and at other U.S. universities through the 1920s and 1930s emphasized the concept of "unit operations," generic processes that underpinned the manufacture of all chemical products. Examples of unit operations included distillation, absorption, filtration, and so forth. These industrial process "building blocks" could, it was believed, be combined and scaled to produce a diverse array of products. But the development of this concept, and greater understanding of the complexities of "scaling up" from laboratory to industrial production volumes, required considerable exposure to industrial practice.

In 1927, the newly established "Development Department" of Standard Oil of New Jersey, which was the nucleus of this firm's industrial research activities, sought the advice of Warren Lewis on ways to exploit the hydrogenation technologies of I.G. Farben, which Standard Oil had obtained through licensing agreements with the German firm. Lewis recommended that the head of MIT's School of Chemical Engineering Practice, Robert Haslam, take a leave of absence to work with Standard Oil. Haslam formed a team of twenty-one researchers from the MIT School that established a research operation at Standard Oil's giant refining complex in Baton Rouge, Louisiana. The Baton Rouge refinery was the site of much of the most important U.S. research in chemical engineering before 1940.

The economic forces that underpinned the restructuring of the petroleum and chemical industries were heavily shaped by technological innovation in other sectors of the economy. The introduction of electric lighting sharply reduced the demand for kerosene, but the rise of the automobile more than offset this decline, supporting as it did the demand for gasoline. Overall demand grew far more rapidly for the lighter products of the oil refinery than for the heavier ones, such as fuel oil. In this important respect the American pattern of demand was substantially different from the European situation, where growth in demand for the lighter products was far lower.

The exploitation of petroleum resources required some essential technological improvements, primarily petroleum "cracking" techniques for breaking large, heavy hydrocarbon molecules into smaller and lighter ones, and, crucially, technologies that would facilitate a shift from batch to continuous-process production. Although William Burton of Standard Oil had developed a technique for the thermal cracking of oil in 1912, widespread adoption of this technique was limited by several deficiencies, including the fact that it could not be operated continuously (Enos 1962, ch. 1). Only in the 1930s and early 1940s did more sophisticated techniques of catalytic cracking, begun by Eugene Houdry of France, establish petroleum refining on a modern basis, leading eventually to the fluid-bed catalytic cracking that is the prevailing refining technology in the world today. Fluid-bed catalytic cracking was the first continuous cracking process technology for high-octane gasoline.

Catalytic cracking technology made possible a much higher degree of control over the output that could be extracted from a given input of petroleum. In practice this meant raising the yield of higher-priced gasoline that could be derived from the heavier crude oil fractions and at the same time capturing the valuable refinery offgases – olefins – as chemical feedstocks. Improvements in the performance of gasolines and associated improvements in engine design were strongly influenced by the U.S. military, which required high-octane aircraft fuel during World War II (Spitz 1988, chs. 2, 3).[57]

The vast expansion in organic chemicals that was triggered by the availability of new feedstocks was powerfully reinforced by the fundamental research in polymer chemistry of German scientists Staudinger, Mark, and Kurt Meyer during the 1920s. Staudinger's research provided a systematic understanding of the structure and behavior of both thermoplastic and thermosetting plastics.[58] Herman Mark, who directed polymer research at I.G. Farben in the 1930s prior to his appointment as a Distinguished Professor of Polymer Science at Brooklyn Polytechnic Institute, noted many years later (in 1976):

Once the basic concepts of this new branch of chemistry were firmly established, polymer chemists settled down to useful and practical work: synthesis of new monomers, quantitative study of the mechanism of polymerization processes in bulk, solution,

---

[57] See Enos (1962) for an authoritative treatment of the development of cracking technologies. Enos's 1958 study underlined the importance of incremental improvements in generating productivity growth in the petroleum refining industry. He studied the introduction of four major new processes in petroleum refining: thermal cracking, polymerization, catalytic cracking, and catalytic reforming. In measuring the benefits flowing from each of these major innovations, he distinguished between the "alpha phase" – or cost reductions that occur when the new process is first introduced – and the "beta phase" – cost reductions that flowed from the later improvements in the new process. Enos found that the average annual cost reductions that were generated by the beta phase of each of these innovations considerably exceeded the average annual cost reductions that were generated by the alpha phase (4.5% as compared to 1.5%). On this basis, he asserted that "the evidence from the petroleum refining industry indicates that improving a process contributes even more to technological progress than does its initial development" (Enos 1958, p. 180).

[58] Thermoplastics are polymers that have the property of softening if heated and of subsequently hardening if cooled. Thermosettings, on the other hand, become permanently rigid if heated.

suspension, and emulsion; characterization of macromolecules in solution on the basis of statistical thermodynamics; study of the fundamentals of the behavior in the solid state. The result was a better understanding of the properties of rubbers, plastics, and fibers. (Spitz 1988, p. 248.)

Mark became an important figure in the introduction of polymer chemistry in the United States. He founded the Institute of Polymer Research at Brooklyn Polytechnic Institute and played a major role in training Americans in polymer chemistry. Many of his students went on to work for Du Pont, the premier American chemical firm, and Mark himself served as a frequent consultant to Du Pont on matters pertaining to polymer research (Hounshell and Smith 1988, pp. 296–297).

International flows of technology remained important in the chemicals industry during the interwar period, particularly within the patent licensing and technology-sharing agreements that linked I.G. Farben, Imperial Chemical Industries, Du Pont, and Standard Oil of New Jersey. In some contrast to more recent "alliances" among firms in technology-intensive industries, which are motivated in part by the desire of participants to expand access to foreign markets, these technology-sharing agreements sought to employ technology exchange in part as a basis for dividing global markets and restricting access by one or another participant to specific areas. The extent of actual technology exchange between the giant German chemicals firm and the U.S. participants in this agreement appears to have been modest. But the technology-exchange agreements linking Imperial Chemical Industries and Du Pont involved more significant bilateral technology flows, especially in the emerging areas of plastics.

The rise of plastics products in the late 1930s initiated the creation of a family of new materials that would eventually replace such conventional materials as glass, leather, wood, steel, aluminum, and paper. Here again, wartime needs accounted for spectacularly

Table 5. Production of Plastic Molding and Extrusion
Materials, 1940–1990.

Year	Thermoplastic Totals, 1000s of lb.	Thermoset Totals, 1000s of lb.
1940	20,300	98,000
1946	239,000	175,000
1950	508,000	286,000
1960	3,785,389	2,126,797
1965	8,448,174	3,236,701
1970	15,685,228	3,524,691
1975	19,728,061	5,139,661
1980	31,121,746	7,064,244
1985	41,755,141	8,242,728
1990	56,754,635	9,500,734

Source: Spitz (1988), p. 229.

rapid growth rates, most particularly in the case of thermoplastics, but rapid growth continued even after the termination of hostilities (Table 5).

Production of plastic materials grew at an average annual rate of more than 13% during the 1945–71 period and declined to an average growth rate of 5.7% per annum during the period 1971–96 (Society of the Plastics Industries, various issues). Rapid growth in production, especially in the early postwar period, was aided by growing use of polyethylene, perhaps the most versatile of all plastics. Polyethylene had been discovered at Imperial Chemical Industries of Great Britain shortly before the Second World War and was used extensively in wartime military applications. The Du Pont Corporation was the first U.S. producer of this product, which it obtained through its licensing agreement with Imperial Chemical Industries. But the rapid growth in U.S. polyethylene output after World War II is attributable in large part to the liberal licensing of polyethylene patents mandated by the U.S. Department of Justice as one of the terms of the settlement of its antitrust suit against Du Pont and Imperial Chemical Industries. In addition, Union Carbide effectively infringed on the Du Pont/Imperial Chemical Industries patents during World War II with the implicit endorsement of the U.S. government and by 1945 had invested in polyethylene production capacity that vastly exceeded that of Du Pont (Smith 1988).

The history of polyethylene's expansion in the U.S. chemicals industry illustrates one of the most important effects of World War II on the U.S. industry, noted by Smith (1988): Under pressure from the U.S. government, and as a result of collaborative production projects during wartime, chemicals process technologies, especially large-scale petrochemicals process technologies, were diffused widely among U.S. firms. The war effectively reduced technology- and patent-based entry barriers within the chemicals industry, and during the postwar period a number of other firms, many of which were oil producers, entered the U.S. industry.

The rapid growth of thermoplastics production in 1940s resulted in part from the entry of new producers, as well as from the arrival of cheap thermoplastics such as polystyrene and polyvinyl chloride.[59] Polyethylene became the largest-volume plastic in the U.S. and world markets after 1945. In 1970, 12 billion pounds of polyethylene were manufactured worldwide, of which the United States produced about 45% (American Chemical Society 1973, p. 68). Low-density polyethylene became the cheapest of all plastics in terms of price per unit.[60] The growing prominence of polymer-

---

[59] Most of these new uses were in rather prosaic applications. As Spitz (1988, p. 229) observed of this period, "producers of refrigerator containers and bags, liners for shipping bags, paper coatings, plastic bottles, and food wrappers were now starting to clamor for this new plastic, which was in extremely short supply."

[60] "In the form of fibres (carpeting, upholstery, sacks, for example), film (bags, wrapping film), rubber goods (tyres, sporting goods), blow-moulded containers (for milk, detergents, cosmetics), uncountable other moulded articles

based products, such as plastics, further reinforced the dependence of the chemical industry on petroleum-based hydrocarbons. This dependence reflected the fact that polymers required the straight-chain (aliphatic) units of petroleum rather than the aromatic rings that predominate in coal tar.

Unlike the technological change connected with innovation in electricity and electronics, it is difficult to identify innovation in the chemical industry with a list of well-known final products such as radios and washing machines. The reasons for the difficulty in identifying new final product chemical inputs point to a central feature of this industry: Most of its output (as much as 75%) consists of intermediate goods that are purchased not by households but by firms in other industries whose products incorporate chemical inputs – paints, fertilizers, pesticides, herbicides, plastics, explosives, synthetic fibers, dyestuffs, solvents, and so forth. Like electricity, therefore, firms in the chemical industry are suppliers of inputs to virtually all sectors of the economy, not the least of which are other firms in the chemical industry itself.

Synthetic Rubber

Another major new product that emerged out of the scientific breakthroughs of polymer chemistry and the crucible of urgent wartime

(housewares, automotive parts, toys, appliances, seating, etc.) and innumerable other items and shapes, stereoregular polymers are touched and used by nearly everyone in one way or another practically every day. The rate of consumption of polypropylene has maintained the heady pace it set from the start and has set the record as the fastest growing polymer the world has yet seen (or appears likely to see). Starting decades behind the others, both polypropylene and linear polyethylene soon joined the select ranks of the 'billion-pound' plastics in the U.S. (shared only with polyvinylchloride, polystyrene, and their predecessor, high-pressure polyethylene)" (McMillan 1979, pp. 169–170).

needs was synthetic rubber. Synthetic rubber has a history long antedating the Second World War in both Germany and the United States. In the United States, Du Pont had introduced a synthetic elastomer, which it named Neoprene, in 1931. Applications of neoprene were limited by its price of $1.05 a pound at a time when natural rubber, which had superior performance characteristics as a general-purpose product, was priced at less than 5 cents per pound.

In 1940, not surprisingly, natural rubber accounted for 99.6% of the U.S. rubber market and synthetic rubber a mere 0.4%. That situation was transformed after December 1941, when Japanese troops overran the plantation sources of natural rubber in southeast Asia. In response, the federal government organized a consortium that initially included the four major rubber companies and Standard Oil. This organization was to pool information concerning styrene-butadiene rubbers and the results of future research. In order to mollify Congressional agricultural supporters, the production of butadiene in the early years of the war relied heavily on alcohol, but by 1945 butadiene, like styrene, had become primarily a petroleum-based product, and petroleum has been the dominant rubber feedstock ever since. The federal government invested approximately $700 million in the construction of fifty-one plants that produced the essential monomer and polymer intermediates needed for the manufacture of synthetic rubber. These facilities were all sold to private firms by the mid-1950s (Morton 1982, pp. 231, 235).

The synthetic rubber program was second only to the Manhattan Project in terms of rapid and extensive mobilization of human resources in order to achieve an urgent wartime goal. By 1945, U.S. consumption of rubber was not only substantially greater than its 1941 level (well over 900,000 long tons versus less than 800,000) but no less than 85% of the 1945 total was accounted for by synthetic rubber (Fig. 3). Synthetic rubber was the first synthetic polymer to be produced in huge quantities from petroleum-based feedstocks (Spitz 1988, p. 141).

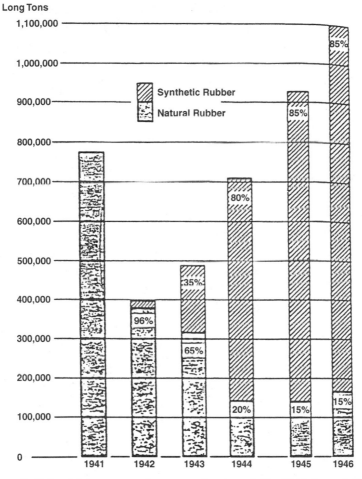

Figure 3. U.S. rubber consumption, natural and synthetic, 1941 through 1946. Synthetic-natural ratio for 1945 and 1946 estimated from 1950–60 data. *Source:* Herbert and Bisio (1985).

The Second World War thus transformed the rubber industry from one that depended on nature for its primary raw material to an industry that depended on a new chemical processing technology for its primary material inputs. The wartime experience of this industry is a compelling illustration of the ability of a technologically

91

dynamic economy such as the United States to overcome natural re-source constraints of an apparently rigid nature. Moreover, this new material basis for the industry illustrated the close technological ties among the chemicals, petroleum, and materials producers that were strengthened further by the development of synthetic fibers.

Synthetic Fibers

In the years immediately following World War II, textile fibers underwent a radical transformation. A number of new synthetic fiber families – mainly polyamides (nylon), acrylics, and polyesters – began to penetrate, and eventually to dominate, markets for the natural products – primarily cotton and wool – that had long underpinned the manufacture and use of textiles. Growth in the use of synthetic fibers took time, because the achievement of optimal fiber characteristics depends on an extensive blending of natural and synthetic materials. Moreover, many of the technical improvements associated with the synthetic materials were eventually transferred to the older natural products, so that by the 1970s the cotton-, wool-, and cellulose-based natural fibers that were used in U.S. textiles and apparel were far different from the fibers that went by those names fifty years earlier. The traditional fiber products eventually acquired such desirable features as ease of cleaning, resistance to wrinkling and shrinking, and flame-retardant finishes, features that were first associated with the new synthetic fibers and were based on the chemical research that had created the synthetic fibers.

Synthetic fibers also shared features with plastics and synthetic rubber, including their origins in the fundamental researches in polymer chemistry of the 1920s and 1930s, involving the work of Staudinger and Mark, as well as that of Wallace Carothers at Du Pont (Hounshell and Smith 1988; Smith and Hounshell 1986). Carothers's research on polymerization yielded neoprene in the late

1920s and culminated in 1935 in the discovery of nylon.[61] Product development in synthetic fibers had begun before the war, primarily at Du Pont and I.G. Farben, but the development of commercial products was disrupted by the voracious needs of the military. Nylon, for example, had made its appearance in women's stockings in 1939, but the new material's high tensile strength and toughness meant that Du Pont's entire output was devoted to military requirements such as parachutes, tires, and tents for the duration of the war.

The new synthetic fibers were based upon monomers that could be derived from coal-based, as well as petrochemical, feedstocks. Cost considerations, however, led to the dominance by petrochemical feedstocks of production of these synthetics, as was the case with plastics and synthetic rubber.[62] Like other postwar synthetics, abundant U.S. petroleum reserves were of central importance to the postwar growth of synthetic fiber production by U.S. firms. But the exploitation of these reserves for synthetics production required considerable effort, and this resource advantage was not a pure gift of nature, as we noted earlier.

As was true of many of the other critical technological advances discussed in this volume, commercialization of initial breakthroughs was extremely time-consuming. A whole range of processing technologies was needed, as well as the development of appropriate methods for producing intermediates, such as terephthalic acid for polyester fibers, or achieving sufficiently high yields of adipic acid for the production of nylon.

Like plastics, the availability of low-cost synthetic fibers led to their use in an expanding array of applications, too numerous to

---

[61] Nylon was initially used as a plastic, replacing hog bristles in toothbrushes in 1937, rather than as a fiber.

[62] Like plastics and synthetic rubber, synthetic fibers were based upon the key petrochemical "building blocks" – ethylene, propylene, butadiene, benzene, and the xylenes. See Fig. 7.3 in Spitz (1988, pp. 298–299).

Table 6. *End-use Markets for Man-Made and*
Synthetic Fibers.

	[Percent of Total U.S. Demand]	
	1954	1966
Women's and children's wear	28	21
Men's and boy's wear	11	11
Home furnishings	12	29
Other consumer uses	10	12
Industrial uses	33	23
Exports	7	4

Source: Spitz (1988), p. 292.

cite individually. By 1968, man-made fibers exceeded (by weight) the combined output of cotton and wool. As of 1966 the leading applications by far were in consumer goods (Table 6). The largest category, accounting for almost one third of the output of synthetic fibers, was in clothing, predominantly for women and children. Home furnishings, including carpeting, drapes, and furniture, were nearly as large. Industrial uses were dominated by tires, followed by reinforced plastics and then by a wide variety of miscellaneous products – hose, rope, belting, bags, filters, and so forth. (American Chemical Society 1973, pp. 89, 95).

Pharmaceuticals

The emergence of the US pharmaceutical industry, an important and distinctive sector of the chemicals industry, drew on roots that were similar to those of the much larger U.S. chemicals industry. The development of both sectors in the mid–19th century United States relied on human skills and competences that originated in

Germany.[63] Not only did the United States depend on imported German pharmaceutical products through the second half of the 19th century, even the most widely used pharmaceutical textbooks were totally dominated by German source materials as late as the 1890s. Moreover, some of the earliest and most successful pharmaceutical manufacturing firms in the United States, such as Pfizer and Merck, were of German origin.

In the course of the 20th century, the American pharmaceutical industry began to exploit a growing domestic stock of scientific knowledge. But the transition to a science-based industry was slow. Well into the 20th century, few new pharmaceutical products could be described as owing their origins to scientific research. This is hardly surprising, because the underlying biomedical disciplines, such as bacteriology, biochemistry, and immunology only began to emerge in the late 19th and early 20th centuries, with the momentous breakthroughs associated with the names of Pasteur, Lister, Koch, and Ehrlich. Many European pharmaceutical firms were affiliates of companies that produced synthetic dyes and fine chemicals. These linkages, however, were lacking in the United States, where pharmaceutical companies remained largely committed to traditional pharmaceutical products (primarily of natural origin) and methods.

Germany and Switzerland dominated world pharmaceutical markets at the outbreak of World War I. Nevertheless, early 20th-century reforms in U.S. medical education and the expansion of medical school curricula to include training in the biomedical

---

[63] Some indication of the extent of that reliance was the very name of the first association in the United States that was devoted to quality assurance in pharmaceuticals. Established in New York in 1851, it bore the name "New Yorker Pharmazeutischer Leseverein," a name that was changed within six months to "Deutscher Pharmazeutischer Leseverein." By the 1850s, many German pharmacists were already receiving university training, a situation very remote from that of the American "frontier" society of midcentury (Feldman and Schreuder 1996).

sciences laid a foundation for future biomedical research (Swann 1988, ch. 2). The passage in 1906 of the Pure Food and Drug Act reflected a growing concern over the sale of pharmaceutical products that were of limited efficacy (or high toxicity) and formed the roots of the much-enlarged federal regulatory presence that emerged later in the 20th century.

World War II initiated a transition in the United States to a pharmaceutical industry that relied on in-house research and eventually, on stronger links with U.S. universities that were also moving to the forefront of research in the biomedical sciences. The surging demand for antibiotics during World War II led to an intensive effort in the United States to exploit Alexander Fleming's discovery of the bactericidal properties of penicillin. Although Fleming's remarkable discovery had been made in 1928, more than a decade later little systematic effort had been mounted to manufacture the drug on a commercial scale.

A massive program to develop technologies for large-scale manufacture of penicillin was orchestrated during World War II by the federal government and involved more than twenty pharmaceutical firms, several universities, and the Department of Agriculture. The success of this "crash program" marked the beginning of a new era of technological change in the U.S. pharmaceutical industry. But the solution to the problems involved in large-scale manufacture of penicillin came not from pharmaceutical chemists but from chemical engineers. These engineers refined the technique of aerobic submerged fermentation, which came to be the dominant production technology, demonstrating its feasibility and improving its yields. The chemical engineers achieved this result by designing and operating a pilot plant in order to solve the complex problems of heat and mass transfer – problems not previously encountered by U.S. pharmaceuticals firms. This joint achievement of the microbiologist and the chemical engineer may be regarded as the first great success of biochemical engineering (American Institute of Chemical Engineers 1970).

Chemicals

The postwar era in the U.S. pharmaceutical industry opened with a widespread expectation in the industry that there existed a vast potential market for new pharmaceutical products, and that catering to this market, however costly, would prove to be highly profitable. These expectations were abundantly fulfilled.

The postwar period also witnessed a remarkable expansion of federal support for biomedical research through the huge growth in the budget of the National Institutes of Health (Fig. 4). Between 1950 and 1965, their budget for biomedical research grew by no less than 18% per year in real terms, and by 1965, the federal government accounted for almost two thirds of all spending on biomedical research. After 1965 this explosive growth rate slowed, but the declining federal share of biomedical R&D spending throughout the 1970s and 1980s also reflected an acceleration in the growth of private R&D funding, especially during the early 1980s. By 1993, total national expenditures for biomedical R&D were more than $30 billion, 39% of which was supported by federal funds and 50% of which was industrially funded (Bond and Glynn 1995, pp. 15–16).

Large pharmaceutical firms, such as Merck, Pfizer, Eli Lilly, and Bristol-Myers, enjoyed rapid growth and high profits during the postwar period. The high profitability of the industry, the source of recurring political controversies and Congressional hearings, was associated with a high level of R&D intensity – on average, company-funded R&D spending accounted for more than 9% of sales among R&D performers in this industry between 1984 and 1994, the highest such level among U.S. manufacturing industries (National Science Foundation 1996, p. 137). During the postwar period, the United States became and has remained the largest source of new pharmaceutical products as well as the largest market for such products. A listing of new products would and does fill a large volume. The major categories include a large number of "antis," beginning with a broad range of antibiotics and going on to include antihypertensives, antiinflammatories, antiulcer drugs, anticholesterols, antidepressants, and antihistamines. It is important

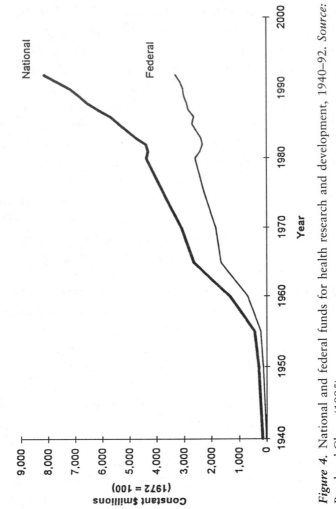

Figure 4. National and federal funds for health research and development, 1940–92. *Source:* Bond and Glynn (1995).

to note that there were no entries in these categories before 1940. Additional categories include vaccines (most notably, that for polio), painkillers, cardiovascular and central nervous system medications, diuretics, vasodilators, oral contraceptives, and alpha- and beta-blockers.

A major discovery in the realm of molecular biology in 1953 eventually set off a new epoch of technological change in the U.S. pharmaceuticals industry. The identification of the double helical structure of DNA by Watson and Crick resulted in more effective methods for drug discovery that gradually replaced the randomized testing that had long dominated the industry (Gambardella 1995). More than forty years after that scientific breakthrough, the biotechnology industry is still in the early stages of its development. A critical step toward a new method for drug creation and manufacture was the gene-splicing technique achieved by Stanley Cohen and Herbert Boyer in 1973, which made possible the alteration of the genetic code of an organism and the manipulation of its subsequent protein production. The new method represented a fundamental discontinuity in the nature of pharmaceutical research, a transition from the realm of chemistry to that of biology. The ongoing revolution in molecular biology has progressed along several trajectories rather than a single paradigm of technological development. Biotechnology has created new techniques for drug discovery, as well as new techniques for production of existing drugs, such as insulin (Henderson, Orsenigo, and Pisano 1998).

The entire biotechnology enterprise has been supported by huge federal expenditures on R&D, including the Nixon Administration's "War on Cancer" of the early 1970s. In the 1980s, the Congressional Office of Technology Assessment estimated that annual federal spending on biotechnology R&D averaged roughly $500 million in the early 1980s and rose to more than $3 billion by fiscal 1990 (Office of Technology Assessment 1984, 1992). Although large investments have been made in the biotechnology industry since the 1970s, a large flow of new products appeared only gradually. Human insulin, the first biotechnology product (based on the use of biotechnological manufacturing processes) to be marketed, received Food and Drug Administration approval in 1982. During the period from 1989 through 1996, the number of publicly traded U.S. companies developing biotechnology-based drugs increased from 45 to 113, and the number of such drugs under development grew from 80 to 284 (Pharmaceutical Research and Manufacturers' Association 1996). By the end of 1996, the FDA had approved thirty-three pharmaceutical products based on biotechnology. In addition, 450 biotechnology-based pharmaceuticals were under development, and more than 120 were in phase III trials (Ernst and Young 1996).

A distinctive feature of the American biotechnology industry has been the prominent role played by new "start-ups,"especially start-ups involving university faculty who act as advisors or entrepreneurs, with financial backing from venture capitalists.[64] But the relationship between the large population of new entrants and the much larger, established pharmaceutical firms has involved a complex new division of labor (Arora and Gambardella 1994), including investments by large pharmaceuticals firms in promising start-up firms, joint ventures, and licensing, and in some cases the acquisition by larger firms of small startups. As Henderson, Orsenigo, and Pisano (1998) have pointed out, start-ups have

---

[64] This is true as well of the broader biotechnology industry within which pharmaceuticals applications have proven to be a lucrative, but by no means the exclusive, focus of new firms. According to one estimate (Ernst and Young 1996), 1,311 firms were active in the U.S. biotechnology industry in 1995; only 20% (265) of these were publicly traded. In addition to pharmaceuticals, the overall biotechnology industry includes diagnostics, agricultural products, and chemical and environmental services. As these data suggest, the industry is still dominated by numerous small start-up companies exploring a wide range of approaches that may lead to new product development. But the contrast with Western Europe is striking – Ernst and Young's annual survey reported that 584 firms, fewer than one half the number present in the U.S. biotechnology industry, were active in biotechnology.

proven to be especially important in applying biotechnology to drug manufacture. The expertise of the established drug firms in organizing and managing clinical trials and other regulatory matters, as well as the established firms' marketing capabilities, means that the biotechnology "research boutiques" often collaborate with established firms, rather than competing directly with them. Applications of biotechnology to the discovery and development of new drugs, however, have been accomplished more successfully by a small number of established pharmaceutical firms with strong links to the academic research community and National Institutes of Health researchers.

The Swiss firm Hoffman La Roche is now the major shareholder in Genentech, perhaps the most successful of the new U.S. biotechnology firms. The large firms serve as repositories of capabilities that are essential to eventual commercial success: extensive distribution networks, marketing savvy, and not least the know-how essential for maneuvering a new pharmaceutical product through a demanding and time-consuming Food and Drug Administration approval process. The shift to a new research paradigm carries with it major implications for industrial structure, firm organization, and specialization. The inability of many traditional pharmaceutical firms to draw on their internal capabilities to make the transition to the new realm of biotechnology has a close parallel in the inability of two of America's most successful chemical firms, Du Pont and Dow, to establish a viable presence in the pharmaceuticals industry (Chandler, Hikino, and Mowery 1998).

5

———

Electric Power

ENTRAL GENERATION OF ELECTRICITY in the United States began
with the opening of the Pearl Street Station in lower Manhattan
in 1882. Although this technology eventually had enormous economic effects, by 1899 electric motors still accounted for less than
5% of total mechanical horsepower in American manufacturing establishments – electric power had not yet had a substantial impact
on the American economy. Indeed, the gradual pace of early adoption of this epochal innovation is yet another example of the gradual realization of the economic impacts of truly major innovations,
reflecting the need for numerous complementary innovations in
technology, organization, and management to support widespread
adoption. In addition, the first version of a new technology of this
type inevitably must be substantially improved through a long series
of incremental innovations and modifications. These modifications
affect both the technology itself and the understanding, on the part
of users, of its potential and operating requirements ("learning by
using" – see Rosenberg [1982]).

The development of electric power generation technologies in the
20th-century United States resembled that of other technological
clusters discussed in this volume in following a path of evolution
that was sensitive to the U.S. natural-resource endowment, even
as it transformed the definition of that endowment. As Hughes
(1983) points out in his history of electric power generation and

transmission technologies, one of the leading sites for the development of hydroelectric-based electric power generation and long-distance transmission was California, abundantly endowed with swift-flowing rivers and narrow (and therefore easily dammed) valleys in the Sierra Nevada. The development of long-distance transmission technologies made these scenic wonders economically valuable sources of power.[65]

A second characteristic feature of the development of electric power generation and transmission technologies in the United States is the close collaboration between U.S. universities and the electric utilities industry in applied research on technical and operating problems. Leading U.S. universities such as Stanford, MIT, and the University of California at Berkeley all developed collaborative research programs with leading U.S. electric utilities and manufacturers of electrical equipment. These programs had far-reaching consequences. In 1929, Vannevar Bush of MIT developed an early version of his differential analyzer, a forerunner of the computer, in a collaborative project with General Electric that sought to simulate the operations of complex electric power grids. At the University of California, collaborative relationships with regional utilities such as Pacific Gas and Electric, as well as their suppliers (e.g., the Federal Telegraph Company of Palo Alto) eventually played an important role in the complex engineering behind Ernest O. Lawrence's cyclotron technologies (Heilbron and Seidel 1989).

---

[65] "The history of electric power in California is a notable reminder that changing technology can make nature a resource, a factor of production. The coming of long-distance power transmission made the remote lakes, streams, and rivers of the Sierra Nevada power sources first for relatively nearby mining towns, then for neighboring farm communities of the great central valley of California, and finally for the far more heavily settled coastal cities" (Hughes 1983, p. 266).

The Growth of Household Electric-Power Consumption

Urban households gained access to electricity in large numbers only between 1910 and 1930. The costs of delivering electricity to rural populations were far greater than for urban residences, and no more than 10% of American farms received electricity from central power stations as late as the early 1930s. The situation in rural areas changed rapidly after federal subsidies were made available through the Rural Electrification Administration, created by President Roosevelt in 1935 (Schurr et al. 1991).

Initially, urban household use of electric power was devoted primarily to lighting. Average residential electricity costs for U.S. households declined from 7.45 cents per kilowatt-hour in 1920 to 6.03 cents per kilowatt-hour in 1930, and residential use of electricity increased more than three-fold in response during this period. The growing availability to consumers of low-cost electric power spawned an expanding array of new products. During the 1920s, the radio, refrigerator, and electric water heater were introduced. The diffusion of these products reflected reductions in the cost of electricity, the greater convenience of this particular form of energy, and the declining costs of products that used electricity, many of which benefited from mass-production manufacturing technologies (Fig. 5). Reductions in the prices (especially quality-adjusted prices) of electricity-using home and office appliances have, if anything, accelerated during the postwar electronics era.

Many of the electrical appliances that became available in mass markets during the 1920s were not fundamentally new. The availability of electricity and the small electric motor breathed life into a number of inventions that had been available, at least in a primitive form, for many years, but that languished because of the absence of an appropriate power source. Such devices as vacuum cleaners, dish-washing machines, and clothes-washing machines

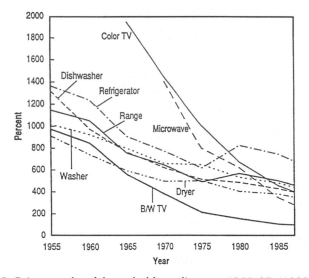

Figure 5. Price trends of household appliances, 1955–87 (1982 dollars). *Sources:* Schurr et al. (1991); Appliance prices: U.S. Bureau of the Census, *Statistical Abstract of the United States, 1969* (Washington, D.C.: GPO, 1969), Table 1157; *1982–1983*, Table 1434; *1988*, Table 1297; GNP deflator: U.S. Council of Economic Advisers, *Economic Report of the President* (Washington, D.C.: GPO, February 1988), Table B-3.

had been developed as far back as the 1850s and 1860s but remained on the shelf until electric motors rendered them practical (Giedion 1948, p. 553). The rate of adoption of electricity-using consumer appliances received an additional impetus from the gradual rise of family incomes during the 1920s.[66] As was true of other mass-produced products, such as the automobile, the adoption of electrical appliances in U.S. households was aided by a more equal distribution of household incomes than prevailed in the contemporary industrial economies of Western Europe.

---

[66] "By the early 1930s, during the Depression, almost all urban homes were wired and had electric irons; 70 percent had radios, and 20 percent to 50 percent had electric refrigerators, washing machines, toasters, vacuum cleaners, and coffee makers" (Schurr et al. 1991, p. 252).

During the post-1945 era, the number and diversity of appliances available for the home increased significantly (Table 7).[67] The consequences of the adoption of this array of home technologies, especially such labor-saving devices as vacuum cleaners and washing machines, for the structure of American life, and even its spatial organization, were profound. Household servants, formerly ubiquitous, became rare in middle-class households, as the quantity of direct labor necessary to maintain a household declined. Labor-saving home appliances also made possible the significant increases in the labor force participation of women that marked World War II and the postwar period. The consequences of electrical appliances do not end with these "minor" changes, however, as the spatial organization of shopping was transformed by the refrigerator's ability to store larger stocks of food for much longer periods of time. No longer were daily or thrice-weekly trips to the grocery store necessary to obtain fresh foodstuffs (Oi 1988). The simultaneous, widespread adoption of the home refrigerator and the automobile made possible the growth of large supermarkets, the displacement of small-scale urban food retailers, and the dispersion of population and retail food purveyors associated with less frequent and longer-distance shopping trips.

Perhaps the most spectacular recent example of growth in the utilization of a new electrical technology, one that uses electric power from batteries rather than from central generation facilities, is the cellular telephone in the 1990s, an innovation that also relied on advances in electronics. The cellular telephone was introduced in 1983, and its developers expected its market to grow slowly. An AT&T prediction at the time projected that cellular telephone

---

[67] The adoption of electric household appliances was promoted by the electric utilities that supplied the energy used by these devices. Throughout the 20th century, until the twin "oil shocks" of 1974 and 1979, the electric-power companies attempted to increase the demand for their product by aggressively marketing electricity-using household devices.

Table 7. *Major Electrical Appliances Introduced in the Postwar Era.*

1950s

Refrigerator-freezer
Television
Clothes dryer
Automatic washing machine
Room air conditioner

1960s

Color television
Dishwasher
Central air conditioning
Space heating
Frost-free refrigerator-freezer
Waste disposal

1970s

Microwave oven
Heat pump
Trash compactor
Food processor

1980s

Home computer
Large-screen television
Video cassette recorder
Compact-disc player
Home satellite receiver

Source: Schurr et al. (1991), Figs. 11.7, 11.8, 11.9.

subscriptions might reach one million by 1999. By the end of 1996, subscriptions had reached 46 million. A major factor in the underestimation of demand was the impact of falling prices and qualitative improvements on the demand for such telephones. In 1983, the average price of cellular phones was roughly $3,000 in current dollars. By 1997, a qualitatively far superior version of this product was available for well under $200 (Hausman 1997). Indeed as of this writing (1998), cellular phones are available for less then $50 when purchased with a telecommunications service contract.

Industrial Applications of Electric Power [68]

No discussion of the impact of electrification on the 20th-century American economy can end with a discussion of the spread of electric appliances within American homes. An equally if not more important trend for economic growth was the adoption of electric power in industrial processes. The reasons for the adoption of electric power, as well as the effects of its industrial application, extend well beyond reductions in energy costs per BTU. The form in which electrically derived energy is delivered was uniquely well-suited to a large number of new industrial technologies, especially in a sector that was fundamental to the development of 20th-century industrial technology: the metallurgical industries.

Electricity in Steel and Aluminum

During the second half of the 19th century, American metallurgy relied primarily on coal as its energy source. During the 20th century, however, one of the most conspicuous aspects of metallurgy has been its growing reliance on electricity as an energy source. The shift from coal to electricity affected virtually every aspect of this

---

[68] See Rosenberg (1998b) for a more detailed discussion of this topic.

sector, ranging from the power of organized labor to the recycling of scrap materials and the nature of the U.S. resource endowment.

The electric furnace had been developed in the late 19th century but was used for only a limited number of specialty steels in which the furnaces produced only a few tons per heat. The use of the electric furnace in these products reflected its freedom from sources of contamination, which was essential to the production of high-quality alloys. As a result, by the early 20th century this technique occupied a small but significant industrial niche in the production of a variety of alloy steels.

During the post-1945 period, continued declines in electricity costs enabled manufacturers to use larger electric furnaces, mainly to produce carbon steel. The minimum efficient size and therefore the capital costs of electric furnaces are far lower than those for the older, traditional steelmaking technologies, and the adoption of the electric furnace spurred the growth of "minimills" in the United States. The vast majority of minimills were founded by new entrants to the steel industry, rather than by the firms operating the integrated steelmills that dominated American industry through most of the 20th century, and the scale of their production facilities was much smaller. In 1985, almost two thirds of the country's mini-mill capacity was in plants that had an annual crude steel capacity of less than 600,000 tons, but by the early 1990s, minimills with production capacity in excess of one million tons were entering operation (Barnett and Crandall 1986, pp. 9–10; Heffernan 1997). As recently as the early 1960s, the use of the electric furnace was essentially confined to sophisticated products such as alloys and stainless steels, and minimills accounted for less than 9% of U.S. raw steel production in 1961 (U.S. Department of Commerce 1975, p. 693). By 1970 this share had grown to over 15%, and by 1994 it constituted nearly 40% (Table 8).

The advantages of electric furnaces are not confined to their cost and size, however; they also can exploit a broader range of raw materials. Electric furnaces commonly operate with a 100% scrap

Table 8. *Share of U.S. Raw Steel Produced in Electric Furnaces, 1970–94.*

Year	Raw Steel Production, *millions of tons*		Electric Furnace Production, *% of Total Production*
	Electric Furnace	Total	
1970	20.2	131.5	15.3
1971	20.9	120.4	17.4
1972	23.7	133.2	17.8
1973	27.8	150.8	18.4
1974	28.7	145.7	19.7
1975	22.7	116.6	19.4
1976	24.6	128.0	19.2
1977	27.9	125.3	22.2
1978	32.2	137.0	23.5
1979	33.9	136.3	24.9
1980	31.2	111.8	27.9
1981	34.1	120.8	28.3
1982	23.2	74.6	31.1
1983	26.6	84.6	31.5
1984	31.4	92.5	33.9
1985	29.9	88.3	33.9
1986	30.4	81.6	37.2
1987	34.0	89.2	38.1
1988	36.8	99.9	36.9
1989	35.2	97.9	35.9
1990	36.9	98.9	37.4
1991	33.8	87.9	38.4
1992	35.3	92.9	38.0
1993	38.5	97.9	39.4
1994	39.6	100.6	39.3

Sources: Barnett and Crandall (1986), p. 7; U.S. Bureau of the Census (1995), p. 776.

charge. The basic oxygen furnace, by contrast, can accommodate up to 50%, but even this amount is uneconomic unless the scrap has been preheated. Consequently, the basic oxygen furnace seldom uses a charge that is more than about one third scrap. The electric furnace has become an attractive way to make relatively inexpensive additions to steelmaking capacity, and it makes possible more intensive exploitation of cheaper inputs than the preexisting technology. The availability of low-cost electricity for iron and steel production has provided a unique opportunity for bypassing the highly energy-intensive earlier stages of mining, coke making, and smelting in conventional steelmaking. If scrap is available, the electric furnace becomes an energy-saving technology, notwithstanding the common but naive complaint that electricity is an "inefficient" technology because of the high thermal losses involved in producing it.

The electric furnace also provides greater locational flexibility than its predecessor technology. It can be located far from coal fields, iron-ore deposits, blast furnaces, or coke ovens, and the introduction of minimills contributed to the decline of steelmaking in western Pennsylvania.[69] Aside from electricity, its main requirement is large "deposits" of urban junk – an input that is, for better or worse, widely available. In fact, the abundance of low cost scrap in the U.S. has been a significant cost advantage to American minimills over their foreign minimill competitors.

The increasing attractiveness of the electric furnace was further aided by an additional characteristic: It is a relatively clean

---

[69] The decentralized geographic distribution of minimills provided another significant cost advantage to the steel manufacturer, because these steel mills rarely were unionized: "Since their plants are scattered around the country, often in small towns in the West and South, their wage rates reflect a variety of local labor-market conditions. Even the largest of the minimills, however, pay wages that are considerably lower than those at the major integrated companies. Total compensation in 1985 for the larger minimills was rarely more than $17.50 per hour, compared with $22.80 for the average integrated company" (Barnett and Crandall 1986, p. 22).

production technology in an industry that had been notorious for its pollution in the past. As federal, state, and local governments have imposed tighter restrictions on emissions, electric furnaces have become more attractive. The air and water pollution of the electric furnace are far easier to deal with than that of the older blast-furnace technology.

Electricity also played a major role in the displacement of open-hearth steelmaking technology by the basic oxygen furnace. Steel industry investors and engineers had long been aware of the usefulness of oxygen in steelmaking – indeed, Bessemer's original patents of the 1850s referred to the possibility of using oxygen in the steel-making process. Although its fundamentals had been understood for a long time, the basic oxygen furnace became commercially feasible only with the availability of cheap oxygen. The technology that was eventually developed for producing pure oxygen, which relies on liquefaction and rectification of air, is highly electricity-intensive. Widespread production and use of industrial oxygen required reductions in electricity costs. A typical plant requires about 350 kilowatt-hours of electricity for each ton of oxygen produced (Shreve and Brink 1977, p. 110). If one adds to the basic oxygen-furnace capacity that of the electric furnaces employed in the U.S. steel industry, it is apparent that an overwhelming fraction of the steel industry's output now depends on electricity – directly in the case of the electric furnace and indirectly in the case of the basic oxygen furnace, which requires large quantities of oxygen that can be economically produced only by an electricity-intensive technology. The open-hearth furnace, which uses far less electric power and accounted for almost 90% of the U.S. steel industry's output in 1959, accounted for a mere 3% in 1989 (Schurr et al. 1991, p. 114).

Aluminum, which became the second most important primary metal in the American economy in the course of the 20th century, is almost inconceivable without the availability of cheap electricity. Although aluminum was first isolated in 1825, it remained little more than a curiosity for a long time. In 1852 it sold for $545 a pound (in 1994 dollars, roughly $7500) – needless to say in very small quantities, and industrial uses were nonexistent. It began its commercial career only after Charles Martin Hall in the United States and Paul Louis Poussaint Heroult in France independently developed an electrolytic process in 1886. The new industry depended on cheap electric power, because huge quantities of electricity were required to separate the aluminum from the oxygen in the ore.[70] These new manufacturing methods became commercially feasible with the availability of cheap electric power at Niagara Falls in the 1890s. Cheap electricity was critical to the widespread production of this new material that was to play such an important role in the 20th century. By making bauxite a commercially attractive raw material for the manufacture of an important industrial material, the application of electricity expanded the U.S. economy's resource base.[71]

Aluminum has been vitally important to 20th-century technologies because it combines high electrical conductivity, high thermal conductivity, and strong resistance to corrosion. Its high strength-to-weight ratio is even more significant because aluminum permits alloying easily and becomes much stronger and stiffer as a consequence. As a result of its combination of light weight and great structural strength, aluminum has come to play a major role in transportation equipment, especially in aircraft. Although the U.S. aircraft industry currently accounts for no more than 0.7% of the output of the U.S. aluminum industry (beverage cans, for both beer and soft drinks, are a much larger market), the contribution of aluminum to aicraft performance is critical.

---

[70] After the bauxite has been converted into aluminum oxide (alumina), the aluminum oxide "is separated into metallic aluminum and oxygen by direct electric current which also provides the heat to keep molten the cryolite bath in which the alumina is dissolved" (Carr 1952, p. 86).

[71] At the same time, cheap electricity gave birth to an entirely new industry, electrochemicals (See Trescott 1981).

Another distinctive feature of aluminum is that it is readily recyclable, and recycling is highly energy-saving. Indeed, recycling of secondary aluminum, which relies on electric-furnace technology, can save up to 95% of the energy consumed in producing aluminum from the original bauxite. Thus, the electric furnace has become the workhorse of the recycling process in the primary metal industries.

Other Industrial Applications of Electricity

Our discussion of the industrial application of electricity so far has focused on the use of this power source in the chemical transformation of materials. For most of the economy, however, electricity has been associated with the introduction of electrically powered machinery. Although it began with the completion of the hydropower complex at Niagara Falls in the last decade of the 19th century, the widespread application of electric power to industry expanded significantly after the turn of the century with the perfection of the steam turbine and the electric motor. From its low level noted at the beginning of this chapter (5% of mechanical horsepower in manufacturing in 1899), industrial use of electric power expanded to more than 25% of the total just ten years later. By 1919 the figure was 55%, by 1929 it was over 82%, and by 1939 it was nearly 90% (Table 9).

Electricity's rise to dominance as a source of industrial power in the U.S. economy was based on its compelling advantages. For one thing, electricity could be packaged in almost any size, whereas steam engines became highly inefficient below a certain size. "Fractionalized" electric-power sources of precisely the right capacity for each industrial application meant large energy and capital savings. Large steam engines generating excessive amounts of power in situations that required only small or intermittent doses no longer were necessary. Electricity thus offered opportunities for "fine tuning" the supply of power to specific needs. Furthermore,

Table 9. *Electric Motor Use as a Fraction of Total Mechanical Horsepower in Manufacturing, Selected Years, 1899–1954.*

Year	Total hp, *1000s*	Electric Motors, *1000s of hp*	Electric Motors, *% of Total hp*
1899	9,811	475	4.8
1904	13,033	1,517	11.6
1909	18,062	4,582	25.4
1914	21,565	8,392	38.9
1919	28,397	15,612	55.0
1925	34,359	25,092	73.0
1929	41,122	33,844	82.3
1939	49,893	44,827	89.8
1954	108,362	91,821	84.7

Source: U.S. Bureau of the Census (1957).

the electric motor reduced requirements for floor space and offered greater freedom in the organization and layout of the workplace. Electric motors meant that the flow of work in factories did not have to accommodate a clumsy system of belts and shafting to transmit power from a central power source to a large number of machines (Du Boff 1967, p. 513).

The benefits of this new technological system in industry, however, took considerable time to be realized. The effects of industrial applications of electric power on measured productivity growth are difficult to detect until the 1920s. The gradual pace of electric power's penetration of industry and productivity gains reflects the high economic and organizational costs of the industrial adoption of this power source. The restructuring of factories, including the reorganization of the flow of work on the factory floor, new work arrangements, and the development of the necessary new patterns of specialization on the part of both workers and management took decades of experimentation and learning (Chandler 1990; David 1990).[72]

The lengthy period of time required for the development of complementary technologies and for the other adjustments that were necessary to realize the full potential of electric power has characterized most major technological innovations in this century. This tendency can be observed not only in electricity-using products but also in the electricity-producing sector itself. Improvements in the production of electric power, like its industrial applications, have relied on a large number of incremental improvements whose development and adoption required decades. The cumulative effect of these numerous small improvements nevertheless was so great that the long-term rate of growth of total factor productivity in this sector was higher than any other sector of the American economy in the first half of the 20th century (Kendrick 1961, pp. 136–137).

Improvements in the efficiency of centralized thermal power plants generated enormous long-term increases in fuel economy. A stream of minor plant improvements, including higher operating temperatures and pressures made possible by new alloy steels and the increases in capacity that have resulted from improved boiler and turbine design, have sharply raised energy output per unit of fuel input. Almost seven pounds of coal were needed to generate a kilowatt-hour of electricity in 1900; production of the same amount of electricity required less than 0.9 pounds of coal in the 1960s (Landsberg and Schurr 1968, pp. 60–61). Even these

---

[72] As David (1990) points out, in the first twenty years of this century electric power was first adopted in new industries that were setting up production facilities for the first time, that is, "producers of tobacco, fabricated metals, transportation equipment and electrical machinery itself." In the older, established industries the introduction of electric power had to await the "physical depreciation of durable factory structures" and the "obsolescence of older-vintage industrial plants sited in urban core areas" (p. 357).

numbers, however, understate the full improvement in the efficiency of energy utilization, in which technological progress in the generation, transmission, and utilization of electric power were all crucial:

> During the 50-year period 1907–1957 reduction of the total energy required or lost in coal mining, in moving the coal from mine to point of utilization, in converting to electric energy, in delivering the electric energy to consumers, and in converting electric energy to end uses have increased by well over 10 times the energy needs supplied by a ton of coal as a natural resource. (U.S. Department of Commerce 1960, p. 501; see also Hughes 1971).

During the 1960s, however, the long trajectory of productivity improvement in electric power generation came to an abrupt end. That end, it is important to note, preceded the sharp increases in energy costs that were associated with the Arab oil boycott and the Iranian revolution during the 1970s. Although the causes of the end of this productivity-growth trajectory are by no means fully understood, it is clear that it contained a large technological component.[73] In particular, the productivity-enhancing possibilities of further expansion in the scale of coal-fired generation plants appear to have been exhausted by the mid-1960s (Gordon 1993). The role of technology in these developments is apparent in trends in thermal efficiency – the amount of fuel required to produce a kilowatt-hour of electricity, which had declined since 1925, ceased its decline in the early 1960s (Fig. 6).

In generating electric power by burning fossil fuels, pressurized steam is produced in a boiler that is passed through a turbine that in

---

[73] The timing of the productivity slowdown in electric power generation raises intriguing questions of its possible connection to the larger issue of the slowdown in overall productivity growth in the American economy that is usually dated from around 1970; see Hirsh (1989), Gordon (1993), and Joskow (1987). This discussion draws on Hirsh (1989) and Joskow (1987); Michael Pries helped to gather pertinent data.

Gains in Thermal Efficiency, 1925-1992

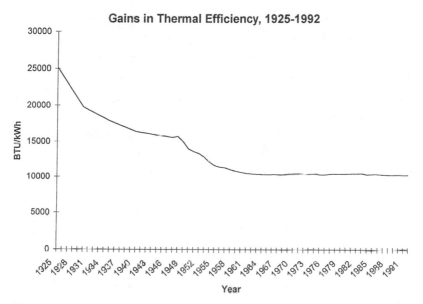

Figure 6. Heat rate for electric power generation in the United States, 1925–92.
Source: Edison Electric Institute (various years).

turn drives an electric generator. The thermal efficiency of this process is an increasing function of steam pressure and temperature.
The story of improved efficiency in electric power generation has
been one of advances in technology that have made possible operation of generating plants at higher temperatures and pressures and
at larger scales. At the beginning of the century turbines operating
at low pressures (180 pounds per square inch) and low temperatures (530°F) produced a mere five megawatts. By the end of the
1920s, boilers were operating at 1400 pounds per square inch and
750°F. The achievement of higher temperatures, in which pulverization of coal played an important role, nevertheless generated
significant side effects, such as the accelerated erosion of furnace
walls. Such erosion was offset and increasing scale achieved by a
combination of new cooling methods and the availability of significantly improved alloys. By the end of the 1930s, boilers were being
operated at 900°F, and by 1953 maximum operating temperatures

119

rose to 1,100°F and maximum pressure to 7500 pounds per square inch. During the postwar period, further improvements in metallurgy and engineering design, many of which had originated during World War II, increased the capacity of the largest fossil fuel plants to over 1,000 megawatts by the 1960s and the average plant to 600 megawatts by 1970.

The thermal efficiency of power plants nonetheless failed to improve above the levels achieved in the early 1960s. Whereas thermal efficiency had increased, on average, from 21.8% in 1948 to 32.2% in 1965, by 1980 it was essentially the same – 32.8% – as it had been fifteen years earlier. The attempt to raise the performance ceilings drastically by recourse to "supercritical" generating units (1200°F and over 4000 pounds per square inch) and larger scale was unsuccessful. Corrosion cracking becomes severe at supercritical levels of pressure and heat and could only be addressed inadequately by incorporating much more expensive alloys. Reliability declined, and higher maintenance and servicing requirements effectively terminated the trajectory of improvements in thermal efficiency that had prevailed since the 1920s.

There is much more to the story: Environmental regulations (e.g., the 1970 Clean Air Act amendments) established emission standards that required scrubbers or substitution of more expensive, low-sulfur fuels. The higher oil prices of the 1970s imposed additional burdens, as well as raising the prospect of costly investments in alternative fuel sources, and contributed to lower levels of utilization of generating capacity. Finally, the enormous federal commitment to nuclear power generation technologies resulted in sharply higher costs for additional generating capacity and serious design, regulatory, and safety problems. Nevertheless, these unfortunate developments followed the end of the steady increases in thermal efficiency in the 1960s, a development that reflects insurmountable obstacles to continued incremental improvements.

Since the 1920s, the share of electrical power in total U.S. energy consumption has grown, although energy intensity for the economy

Producer Price Index/GDP Deflator

Figure 7. Electricity producer price index/GDP deflator for the United States, 1959–95. *Source:* U.S. Dept. of Commerce, Bureau of Economic Analysis (1996).

as a whole, or the ratio of energy consumption to gross national product, has declined markedly. These trends were connected, until the late 1960s, by a decline in the relative price of electricity – Fig. 7 displays longitudinal trends in the deflator for electricity prices relative to that for gross domestic product. The relative price of electricity began to rise beginning in the late 1960s and continued to do so until the mid-1980s. Although the relative price of electricity has declined in subsequent years, it remains substantially above its level of the late 1960s.

6

The Electronics Revolution, 1947–90

LIKE MOST OF THE MAJOR TECHNOLOGICAL advances considered in this volume, electricity and its associated innovations were complex systems of technologies, advances that frequently relied heavily on incremental improvements in individual components. Its complex and "systemic" nature meant that both adoption and realization of the productivity-enhancing effects of electrification took considerable time. An important characteristic of the evolution of electrical technologies, as well as chemicals and the internal combustion engine, is the frequent appearance of "technology bottlenecks," often centered around individual components or the interconnections of components, within the system. Such bottlenecks also launched and guided the evolution of electronics technologies. The emergence of a critical bottleneck in telecommunications, as we note in this chapter, motivated Bell Telephone Laboratories to undertake the research program that produced the first transistors and ultimately launched the postwar electronics revolution. The subsequent development of electronics components and the computer systems into which they are incorporated has been influenced by the enduring need to resolve obstacles to further progress that are imposed by other elements of these complex systems – examples include excessive numbers of discrete components, complex software, and a lack of interchangeability in components.

Advances in electronics technology created three new industries – electronic computers, computer software, and semiconductor components – in the postwar U.S. economy. Electronics-based in-novations supported the growth of new firms in these industries and revolutionized the operations and technologies of more mature in-dustries, such as telecommunications, banking, and airline and rail-way transportation. The electronics revolution can be traced to two key innovations – the transistor and the computer. Both appeared in the late 1940s, and the exploitation of both was spurred by Cold War concerns over national security. The creation of these innova-tions also relied on domestic U.S. science and invention to a greater extent than many of the critical innovations of the pre-1940 era.

Semiconductors

The transistor was invented at Bell Telephone Laboratories in late 1947 and marked one of the first tangible payoffs to an ambi-tious program of basic research in solid-state physics that Mervin Kelly, Bell Laboratories' director, had launched in the 1930s. Fac-ing increasing demands for long-distance telephone service, AT&T sought a substitute for the repeaters and relays that would other-wise have to be employed in huge numbers, greatly increasing the complexity of network maintenance and reducing reliability. Kelly felt that basic research in the emergent field of solid-state physics might yield technologies for this purpose.[74]

Commercial exploitation of Bell Laboratories' discovery was influenced by U.S. antitrust policy, cited earlier as an important

[74] "As early as 1936, Kelly felt that one day the mechanical relays in telephone exchanges would have to be replaced by electronic connections because of the growing complexity of the telephone system and because much greater demands would be made on it. As this is hardly technically feasible using valves, it seems that Kelly was thinking not simply of a radically new valve technology, but perhaps of radically new electronics . . . It seems most likely that Kelly saw the logical progression from a semiconductor rectifier in copper oxide to be a semiconductor switch" (Braun and MacDonald 1982, p. 36).

influence on the evolution of the overall U.S. R&D system throughout this century. In 1949, the U.S. Department of Justice filed a
major antitrust suit against AT&T. Faced with this threat to its
existence, AT&T was reluctant to develop an entirely new line of
business in the commercial sale of transistor products and may have
wished to avoid any practice that would draw attention to its market power, such as charging high prices for transistor components or
patent licenses. In April 1952, Bell Laboratories held a symposium
open to all (for a $25,000 admission fee) that revealed the technol-
ogy of the point-contact transistor and explained progress in the
manufacture of junction transistors (Brooks 1976, p. 54). In 1956,
the antitrust suit was settled through a consent decree, and AT&T
restricted its commercial activities to telecommunications service
and equipment. The 1956 consent decree also led AT&T, holder of
a dominant patent position in semiconductor technology, to license
its semiconductor patents at nominal rates to all comers, seeking
cross-licenses in exchange for access to its patents. As a result, virtually every important technological development in the industry
was accessible to AT&T and all of the patents in the industry were
linked through cross-licenses with AT&T.

The first commercially successful transistor was produced by
Texas Instruments, rather than by AT&T, in 1954. Moreover, like
the other major innovations discussed in this volume, the Texas
Instruments transistor was a major modification of the original
Bell Laboratories device; the design changes lowered the costs of
fabrication and improved reliability. The development of Texas Instruments' junction transistor required extensive incremental improvements in the fabrication and purification of silicon, as well
as advances in device design. The silicon junction transistor was
quickly adopted by the U.S. military for use in radar and missile
applications.

The next major advance in semiconductor electronics was the
integrated circuit (IC), which combined a number of transistors on
a single silicon chip, in 1958. The IC was in large part a response

to the growing reliability problems associated with systems that utilized large numbers of discrete transistors. As the number of transistors employed in a system grew, the probability that the failure of a single component or interconnection would cause a failure in the system increased exponentially.[75] Continued growth in demand for semiconductor components required a new class of products whose price and features (e.g., greater reliability and fewer interconnections) would expand application opportunities in systems. The IC was invented by Jack Kilby of Texas Instruments and drew on process innovations in diffusion and oxide-masking technologies that had initially been developed for the manufacture of silicon junction transistors. The development of the IC made possible the interconnection of large numbers of transistors on a single device, and its commercial introduction in 1961 spurred growth in industry shipments (Fig. 8 displays trends in the composition of industry shipments between 1955 and 1990).

Kilby's search for the IC was motivated by the perceived desirability of a device that could expand the military (and, eventually, the commercial) market for semiconductor devices.[76] Little of Kilby's pathbreaking R&D was supported by the U.S. military; the military's greatest contribution to the early development of the

---

[75] "As long as each element had to be made, tested, packed, shipped, unpacked, retested and interconnected with others, it would be sheer individuality of components rather than technical or production limitations which would constrain improvement. The problem posed by the interconnection of components was particularly severe for, no matter how reliable the components, they were ultimately only as reliable as the joints connecting them and the generally manual methods used for wiring circuits. The more complex the system, the more interconnections were needed and the greater the chance of failure through this cause. Hence, the main obstacle to progress was a tyranny of numbers" (Braun and MacDonald 1982, p. 99).

[76] From the beginning, Texas Instruments and other firms were aware of the commercial potential of the IC. As one of its first demonstration projects, Texas Instruments constructed a computer to demonstrate the reductions in component count and size that were possible with ICs.

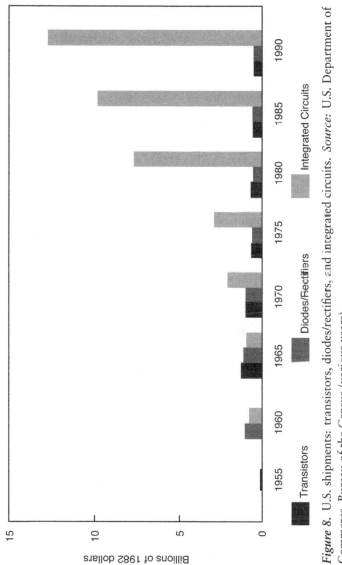

Figure 8. U.S. shipments: transistors, diodes/rectifiers, and integrated circuits. *Source:* U.S. Department of Commerce, Bureau of the Census (various years).

IC industry was its demand for highly reliable components. The "tyranny of numbers" problem of discrete components was especially acute in computer applications in military systems (the Minuteman missile guidance system, for example, was a rugged, high-performance computer). The demands of military computer designers for high reliability and ruggedness in components ensured that these systems would offer the first opportunities to apply ICs.[77]

Once military and space systems demonstrated the viability of the IC, commercial computer applications quickly emerged for the new technology. Table 10 shows the percentage of discrete semiconductor production used in military and space applications uses between 1955 and 1968 (including the National Aeronautics and Space Administration (NASA), the Federal Aviation Administration (FAA), and the Atomic Energy Commission (AEC)). Commercial demand for discrete semiconductors was also large in the early years of the industry, as these components were used in inexpensive hearing aids and radios that tapped a mass market. Although military demand for discrete semiconductors peaked during the 1960–62 Minuteman missile program and increased again with the Vietnam War buildup of the mid-1960s, defense demand declined as a proportion of output throughout the 1960s.

ICs overtook transistors in sales by 1966, and the use of ICs in electronic systems (e.g., computers) began to restructure the demand for other semiconductor components. By the mid-1970s, non-IC semiconductors were used in most systems applications as complements to ICs, and demand growth for non-IC components therefore depended on the growth of markets for ICs. Figures 9 and 10 show the growth in total IC production and changes in the mix of IC products between 1972 and 1990. The value of total IC shipments

---

[77] "It was said that if all military components received the cosseting given to those in Minuteman, the expense would have exceeded the gross national product" (Braun and MacDonald 1982, p. 99).

Table 10. U.S. Production of Semiconductors for Defense
Requirements, 1955–68.

Year	Total Semiconductor Production, millions of dollars	Defense Semiconductor Production,* millions of dollars	Production for Defense, % of Total
1955	40	15	38
1956	90	32	36
1957	151	54	36
1958	210	81	39
1959	396	180	45
1960	542	258	48
1961	565	222	39
1962	575	223	39
1963	610	211	35
1964	676	192	28
1965	884	247	28
1966	1,123	298	27
1967	1,107	303	27
1968	1,159	294	25

* Defense production includes devices produced for the Department of Defense, Atomic Energy Commission, Central Intelligence Agency, Federal Aviation Agency, and National Aeronautics and Space Administration.
Sources: Electronic Industries Association, Electronic Industries Yearbook: 1969 (Washington, DC: Electronic Industries Association, 1969); U.S. Department of Commerce, Business and Defense Services Administration, Electronic Components: Production and Related Data, 1952–1959 (Washington, DC: U.S. Government Printing Office, 1960); U.S. Bureau of the Census, Current Industrial Reports: Semiconductors, Printed Circuit Boards, and Other Electronic Components, various years (Washington, DC: U.S. Government Printing Office).

grew by more than 20% annually during this period. Rapid growth in output was accompanied by significant changes in its composition. The microprocessor, invented in 1971, accounted for $275 million in revenue by 1976 (included in the "metallic oxide silicon" category); revenues from older IC product classes, such as diode

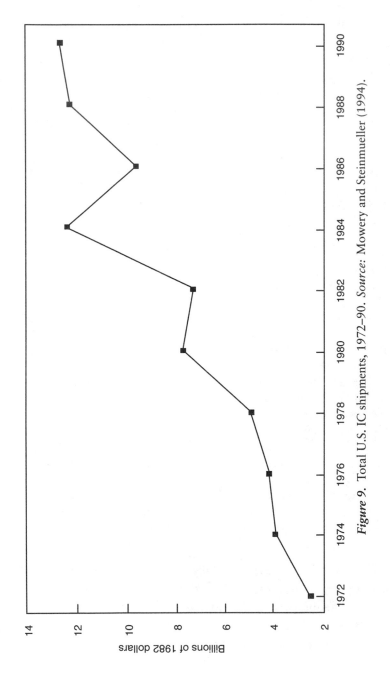

Figure 9. Total U.S. IC shipments, 1972–90. *Source:* Mowery and Steinmueller (1994).

130

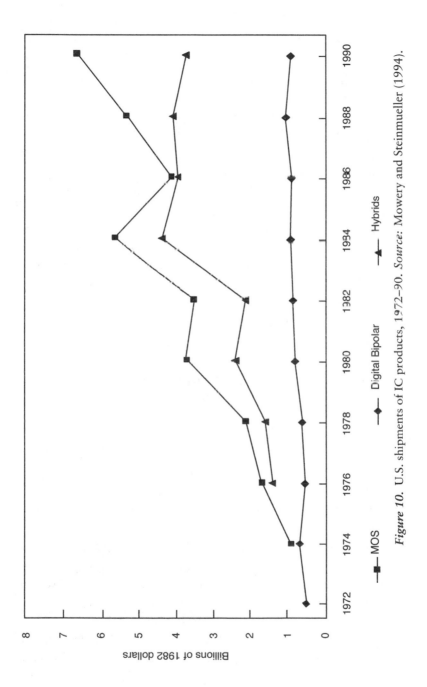

Figure 10. U.S. shipments of IC products, 1972–90. *Source:* Mowery and Steinmueller (1994).

transistor logic (included in "digital bipolar" in Fig. 10), began to fall in the late 1970s.

One result of the high level of federal government involvement in the early postwar semiconductor industry, as both a funder of R&D and a purchaser of its products, was the emergence of a structure for the innovation and technology commercialization processes that contrasted with that of pre-1940 technology-intensive U.S. industries, such as chemicals or electrical machinery. In a virtual reversal of the prewar situation, the R&D facilities of large firms provided many of the basic technological advances that new, smaller firms commercialized. Small entrants' role in the introduction of new products, reflected in their often-dominant share of markets in new semiconductor devices, significantly outstripped that of larger firms. Moreover, the role of new firms grew in importance with the development of the integrated circuit.

In 1960, just prior to the commercial introduction of the IC, the established producers of electronic systems, most of which were founded before 1940 and entered the electronics industry from the office equipment, consumer products, or electrical machinery industries, accounted for five of the ten largest U.S. manufacturers of transistors. By 1975, however, the dominant producers in this new industry included many more relatively new firms, such as Intel and Fairchild, that had entered the industry in the late 1950s and had grown rapidly by exploiting their expertise in integrated circuits. Only two of the five electronic-system firms that had been among the ten largest producers of transistors remained in the ranks of the top ten producers of semiconductors by 1975 (Levin 1982, p. 30). The IC, much more than the transistor, transformed the structure of the U.S. semiconductor industry, and the new firms that emerged as leaders sold the vast majority of their output to other firms, rather than producing primarily for internal consumption.

Although the military market for ICs was rapidly overtaken by commercial demand, military demand spurred early industry growth and price reductions that eventually would create a large

Table 11. U.S. Integrated-Circuit Production and Prices, and the
Importance of the Defense Market, 1962–68.

Year	Total Production, millions of dollars	Average Price per Integrated Circuit, dollars	Defense Production Share of Total Production*, %
1962	4[i]	50.00[†]	100[†]
1963	16	31.60	94[†]
1964	41	18.50	85[†]
1965	79	8.33	72
1966	148	5.05	53
1967	228	3.32	43
1968	312	2.33	37

* Defense production includes devices produced for Department of Defense, Atomic Energy Commission, Central Intelligence Agency, Federal Aviation Agency, and National Aeronautics and Space Administration.
[†] Estimated.
Sources: Electronic Industries Association, *Electronic Industries Yearbook: 1969* (Washington, DC: Electronic Industries Association, 1969); U.S. Department of Commerce, Business and Defense Services Administration, *Electronic Components: Production and Related Data, 1952–1959* (Washington, DC: U.S. Government Printing Office, 1960); U.S. Bureau of the Census, *Current Industrial Reports: Semi-conductors, Printed Circuit Boards, and Other Electronic Components*, various years (Washington, DC: U.S. Government Printing Office).

commercial market for ICs (Table 11). Military procurement policies also influenced industry structure. In contrast to Western European defense ministries, the U.S. military was willing to award substantial procurement contracts to firms, such as Texas Instruments, that had recently entered the semiconductor industry and that had little or no history of supplying the military.[78] Interestingly,

[78] "European governments provided only limited funds to support the development of both electronic component and computer technology in the 1950s and were

133

however, military R&D contracts during this period had only a modest influence on innovation. The major corporate recipients of military R&D contracts were not among the pioneers in the introduction of innovations in semiconductor technology; the pioneering firms did so without military R&D contracts (Kleiman 1966, pp. 173–174).

The U.S. military's willingness to purchase from untried suppliers was accompanied by conditions that effectively mandated substantial technology transfer and exchange among U.S. semiconductor firms. To reduce the risk that a system designed around a particular IC would be delayed by production problems or by the exit of a supplier, the military required its suppliers to develop a "second source" for the product, a domestic producer that could manufacture an electronically and functionally identical product. To comply with second-source requirements, firms exchanged designs and shared sufficient process knowledge to ensure that the component produced by a second source was identical to the original product.

By facilitating entry and supporting high levels of technology spillovers among firms (e.g., the 1956 AT&T consent decree, the Department of Defense second-source policy), public policy and other influences increased the diversity and number of technological alternatives explored by individuals and firms within the U.S. semiconductor industry during a period of significant uncertainty about the direction of future development of this technology (for a discussion of the role of uncertainty in technological change, see Rosenberg [1996]). Extensive entry and rapid interfirm technology

---

reluctant to purchase new and untried technology for use in their military and other systems. European governments also concentrated their limited support on defense-oriented engineering and electronics firms. The American practice was to support military technology projects undertaken by industrial and business equipment firms that were mainly interested in commercial markets. These firms viewed their military business as a development vehicle for technology that eventually would be adapted and sold in the open marketplace" (Flamm 1988, p. 134).

diffusion also fed intense competition among U.S. firms. The intensely competitive industry structure and conduct enforced a rigorous "selection environment," ruthlessly weeding out less effective firms and technical solutions. For a nation that was pioneering in the semiconductor industry, this combination of technological diversity and strong selection pressures proved to be highly effective.

In some contrast to their prominence in the development of the chemical industry or the later development of the U.S. computer software industry, U.S. universities played a minor role as direct sources of the technologies applied in the emergent semiconductor industry. The reasons for this are unclear, although the extraordinarily complex nature of the manufacturing processes involved in the industry may have made it impossible for university-based researchers to replicate the process technologies necessary to contribute to industrial practice. Even the origins of the solid-state physics theory that Shockley and colleagues applied so brilliantly at Bell Laboratories lay as much within Bell Laboratories as within academia; the first widely used textbook, *Electrons and Holes in Semiconductors*, was written by Shockley. But U.S. universities were quick to develop courses and graduate programs of study to train the engineers and scientists who were needed by this industry. U.S. universities were aided in this task by substantial research funding from the federal government, much of which was defense-related.

The Computer*

The development of the U.S. computer industry also benefited from Cold War military spending, but in other respects the origins and early years of this industry differed from semiconductors. Although they were at best peripheral actors in the early development of

---

* This section draws on Langlois and Mowery (1996).

semiconductor technology, U.S. universities were important sites for the early development activities, as well as the research, that led to the earliest U.S. computers. Federal spending during the late 1950s and 1960s from military and nonmilitary sources provided an important basic research and educational infrastructure for the development of this new industry.

During the war years, the American military sponsored a number of projects to develop high-speed calculators to solve special military problems. The ENIAC – generally considered the first fully electronic digital computer – was funded by Army Ordnance, which was concerned with the computation of firing tables for artillery. Developed by J. Presper Eckert and John W. Mauchly at the Moore School of the University of Pennsylvania, the ENIAC did not rely on software but was hard-wired to solve a particular set of problems. In 1944, John von Neumann began advising the Eckert-Mauchly team, and his innovations were reflected in the architecture of their next machine, the EDVAC, which was the first stored-program computer. Instead of being hard-wired, the EDVAC's instructions were stored in memory, facilitating their modification.

Von Neumann's abstract discussion of the concept (von Neumann 1945) circulated widely and served as the logical basis for virtually all subsequent computers.[79] But even after the von Neumann scheme became dominant, which occurred rapidly in the 1950s, software remained closely bound to hardware. During the early

[79] Like the semiconductor industry, but for different reasons, intellectual property rights were relatively weak in the early years of the computer industry. One reason for this was the extensive dissemination of the EDVAC report, which led Army patent lawyers to rule that "because of the time elapsed since publication of the EDVAC report [Eckert/Mauchly/von Neumann], the concepts related to EDVAC-type machines were in the public domain. Other groups would use these ideas in designing their computers over the next few years" (Flamm 1988, p. 50). The subsequent settlement in 1956 of a federal antitrust suit against IBM also included liberal licensing decrees, further supporting liberal interfirm diffusion of computer technology.

1950s, the organization designing the hardware generally designed the software as well. As computer technology developed and the market for its applications expanded after 1970, however, users, independent developers, and computer service firms began to play prominent roles in software development.

Although military support for the ENIAC and other projects began with narrowly defined goals, these programs produced general principles and technologies that soon found much broader application. Indeed, in the case of Whirlwind (Redmond and Smith 1980), the Navy never obtained its hoped-for flight simulator. The Whirlwind project was by far the most expensive of the early postwar federal computer programs[80] and was spared only when the U.S. Air Force adopted it as the basis for the SAGE air-defense program that began in the early 1950s. In addition to driving the development of a reliable large computing system and the communications technologies necessary to link these computers with radar networks, SAGE was among the earliest programs in large-scale software development (Tropp 1983).

The first fully operational stored-program computer in the United States was the SEAC, a machine built on a shoestring by the National Bureau of Standards in 1950 (Flamm 1988, p. 74). A number of other important machines were developed for or initially sold to federal agencies. Among them were

- The IAS computer, 1951, built by von Neumann at the Institute for Advanced Study on the basis of his EDVAC and subsequent papers. Funding came from the Army, the Navy, and RCA, among others.
- The Whirlwind, 1949, developed at MIT and the source of advances in computer technologies that were incorporated into the SAGE strategic air-defense system of the 1950s.

[80] The Whirlwind's cost of $3 million substantially exceeded the average cost of $650,000 for the other systems described here (Redmond and Smith 1980).

- UNIVAC, 1953, built by Remington Rand, which had bought the rights to the Eckert-Mauchly technology. Early customers included the Census Bureau and other government agencies as well as private firms.
- The IBM 701, 1953, developed by IBM and influenced by the IAS design. This computer was developed as a scientific computer for the Defense Department, which bought most of the first units.

From the earliest days of their support for the development of computer technology, the U.S. armed forces were anxious that technical information on this innovation reach the widest possible audience. This attitude, which contrasted with that of the military in Great Britain or the Soviet Union, appears to have stemmed from the U.S. military's concern that a substantial industry and research infrastructure would be required for the development and exploitation of computer technology.[81] The technical plans for the military-sponsored IAS computer were widely circulated among U.S. government and academic research institutes and spawned a number of "clones" (e.g., the ILLIAC, the MANIAC, AVIDAC, ORACLE, and JOHNIAC – see Flamm [1988, p. 52]).

Although much of the Navy's cryptology-related research in computer technology remained classified, the Office of Naval Research

---

[81] Goldstine (1972, p. 217), one of the leaders of the wartime project sponsored by the Army's Ballistics Research Laboratory at the University of Pennsylvania that resulted in the Eckert-Mauchly computer, notes that "A meeting was held in the fall of 1945 at the Ballistic Research Laboratory to consider the computing needs of that laboratory 'in the light of its post-war research program.' The minutes indicate a very great desire at this time on the part of the leaders there to make their work widely available. 'It was accordingly proposed that as soon as the ENIAC was successfully working, its logical and operational characteristics be completely declassified and sufficient publicity be given to the machine ... that those who are interested ... will be allowed to know all details.'" Goldstine is quoting the "Minutes, Meeting on Computing Methods and Devices at Ballistic Research Laboratory," 15 October 1945.

organized seminars on automatic programming in 1951, 1954, and 1956 (Rees 1982, p. 120). Along with similar conferences sponsored by computer firms, universities, and the meetings of the fledgling Association for Computing Machinery (ACM), the Office of Naval Research (ONR) conferences circulated ideas within a developing community of practitioners that did not yet have journals or other formal channels of communication (Hopper 1981). The Institute for Numerical Analysis at UCLA, established with support from the Office of Naval Research and the National Bureau of Standards (Rees 1982, pp. 110–111), made important contributions to the overall field of computer science.

As of 1954, the ranks of the largest U.S. computer manufacturers were dominated by established firms in the office equipment and consumer electronics industries. The group included RCA, Sperry Rand (originally the typewriter producer Remington Rand, which had acquired Eckert's and Mauchly's embryonic computer firm), and IBM, as well as Bendix Aviation, which had acquired the computer operations of Northrop Aircraft. Sales of computers by these firms went primarily to federal government agencies, particularly the defense and intelligence agencies.

Business demand for computers gradually expanded during the early 1950s to form a substantial market. The most commercially successful machine of the decade, with sales of 1,800 units, was the low-priced IBM 650 (Fisher et al. 1983, p. 17). The 650, often called the Model T of computing, thrust IBM into industry leadership (Katz and Phillips 1982, p. 178; Flamm 1988, p. 83). Even in the case of the 650, however, government procurement was crucial: The projected sale of fifty machines to the federal government (a substantial portion of the total forecast sales of 250 machines) influenced IBM's decision to initiate the project (Flamm 1988).

Programming all of these early machines was a tedious process that resembled programming a mechanical calculator: The programmer had to explicitly specify in hardware terms (the memory addresses) the sequence of steps the computer would undertake. This characteristic tied software development closely to a particular machine, because programmers had to understand its hardware architecture. Because few models of any single machine were available, programming techniques developed for one machine had very limited applicability. This factor made the commercial success of the IBM 650 crucial to advances in software and in programming techniques; the 650 created a generic "platform" for the development of programs that could run on a large installed base.[82] The large commercial market for computers that was created by the 650 provided strong incentives for industry to develop software for this architecture.

University research played a key role in the growth of the U.S. computer industry. Universities were important sites for applied, as well as basic, research in hardware and software and contributed to the development of new hardware. In addition, of course, the training by universities of engineers and scientists active in the computer industry was extremely important. By virtue of their relatively "open" research and operating environment that emphasized publication, relatively high levels of turnover among research staff, and the production of graduates who sought employment elsewhere, universities served as sites for the dissemination and diffusion of innovations throughout the industry.

U.S. universities provided important channels for cross-fertilization and information exchange between industry and academia, but also between defense and civilian research efforts in software and

---

[82] "Prior to this system [the IBM 650], universities built their own machines, either as copies of someone else's or as novel devices. After the 650, this was no longer true. By December 1955, Weik reports, 120 were in operation, and 750 were on order. For the first time, a large group of machine users had more or less identical systems. This had a most profound effect on programming and programmers. The existence of a very large community now made it possible, and indeed, desirable, to have common programs, programming techniques, etc." (Goldstine 1972, p. 331).

in computer science generally. Hendry (1989) argues that a lack of interchange between military and civilian researchers and engineers weakened the early postwar British computer industry;[83] the very different situation in the U.S. enhanced the competitiveness of this nation's hardware and software industry complex. The smaller role of universities in computer science and software-related research activities in Japan and the Soviet Union also reduced the flow of knowledge among different research sites and hampered the pace of technological progress in these nations' software industries.

The private sector took some of the first steps to begin building the discipline of computer science within U.S. universities. In addition to price discounts on its machines, Control Data Corporation offered research grants, free computer time, and cash contributions to U.S. universities (Fisher, McKie, and Mancke 1983, p. 170). IBM donated computer time to establish regional computing centers at MIT and UCLA in the mid-1950s,[84] and rented some 50 of its model 650 computers to universities at reduced rates[85]

---

[83] "Indeed, despite what was in many respects a first-rate network of contacts, the NRDC [National Research and Development Corporation] was not even aware of some of the military computer developments taking place in the 1950s and early 1960s. Nor were the people carrying out these developments in many cases aware of work on the commercial front. In America, in contrast, communications between different firms and laboratories appear to have been very good, even where classified work was involved" (Hendry 1989, p. 162).

[84] In the case of MIT, IBM donated a model 704 computer in 1957, which was available free of charge to MIT seven hours a day and to twenty-four other New England universities another seven hours a day. IBM itself used the remaining ten (nighttime) hours (Wildes and Lindgren 1985, pp. 336–337).

[85] The IBM educational allowance program began in October 1955, with 60% reductions in lease rates to universities. In May 1960, IBM changed the allowance to 20% for administrative use and 60% for academic use. In 1963, the company abandoned the administrative/academic distinction and reduced all allowances to 20% on new orders. In 1965, IBM set up a sliding scale of allowances on the new 360 series, ranging from 20% on the base model to 45% on a high end system. By 1969, the allowance had been reduced to 10% (Fisher, McKie, and Mancke 1983, p. 172).

(Galler 1986; Fisher, McKie, and Mancke 1983, pp. 170–172). Computer manufacturers recognized that in addition to the public-relations benefits of supporting higher education, they could increase demand for their products by facilitating the acquisition and use of their hardware at universities (Fisher, McKie, and Mancke 1983, p. 169). Support of academic computing would attack the already apparent software "bottleneck" by training more programmers and might also "lock in" future users and buyers of computer equipment to a firm's proprietary design or architecture.[86]

Federal policy also aided the central role of U.S. research universities in the advance of hardware and software technologies. Even after the rise of a substantial private industry dedicated to the development and manufacture of computer hardware, federal R&D support aided the creation of the new academic discipline of computer science. The institution-building efforts of the National Science Foundation and the Defense Department came to overshadow private-sector contributions by the late 1950s. Figure 11 depicts the growth in constant-dollar National Science Foundation expenditures on computer science research, and Fig. 12 points out the important role played by the Defense Advanced Research Projects Agency in the growth of federal support for computer science research in U.S. universities. In 1963, about half of the $97 million spent by universities on computer equipment came from the federal government; the universities themselves paid for 34% and computer makers picked up the remaining 16% (Fisher, McKie, and Mancke 1983, p. 169).

The federal government's expanding role in supporting R&D, much of which was located in U.S. universities, during the 1950s

---

[86] "The grants were in IBM's interest, because the corporation felt a strong concern with supporting and maintaining a close relationship with universities, and because an entire generation of students and faculty would associate computers and computing with 'IBM'" (Galler 1986, p. 37).

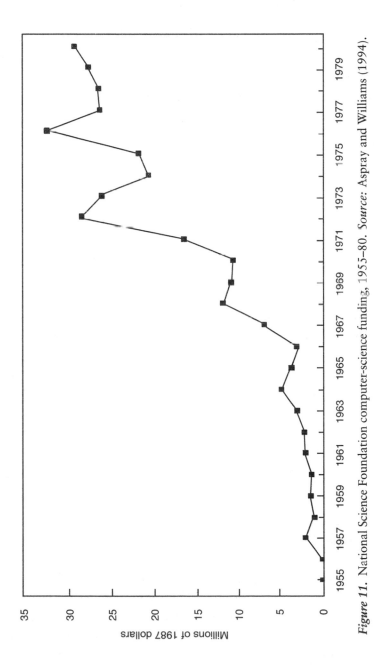

Figure 11. National Science Foundation computer-science funding, 1955–80. *Source:* Aspray and Williams (1994).

143

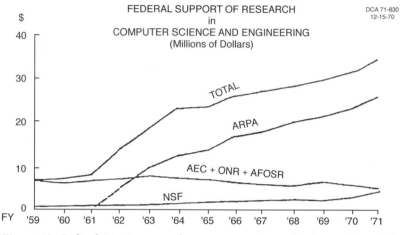

Figure 12. Federal R&D support for computer science, fiscal years 1959–71. *Source:* National Science Foundation (1970).

was supplemented by procurement spending on military systems. In both the hardware and software areas, the government's needs differed from those of the commercial sector, and the magnitude of purely technological spillovers from military R&D and procurement to civilian applications appear to have declined somewhat as the computer industry moved into the 1960s. As was the case in semiconductors, however, military procurement demand acted as a powerful attraction for new firms to enter the industry, and many such enterprises entered the fledgling U.S. computer industry in the late 1950s and 1960s. The most conspicuous early example of defense-related computer development and procurement was the SAGE air-defense system, the computerized early-warning system developed and deployed in the 1950s, which involved what was by far the largest programming effort of the day. In 1950, the Air Force established the MIT Lincoln Laboratories to develop air-defense technology to protect the continental United States against a strategic bomber attack. This effort absorbed MIT's Whirlwind project and evolved into SAGE, the Semi-Automatic Ground Environment.

Successful tests of the SAGE system on Cape Cod led to a full-scale development effort in 1953, coordinated by Lincoln Labs. Lincoln Labs chose IBM to produce computers that were based on the Whirlwind model; AT&T developed the communications system that linked the radar units; and Burroughs built peripheral equipment. A division of the Rand Corporation that soon spun off to become System Development Corporation took up the massive programming task.[87]

The progress of computer technology since the 1950s has been driven by the interaction of several trends: dramatic declines in the price-performance ratios of components, including central processing units and such essential peripherals as data storage devices; resulting in part from those declines, the rapid extension of computing technology into new applications; and the increasing relative costs of software. These trends have created bottlenecks that have influenced the path of technological change. The IBM 360 mainframe computer, for example, which cemented IBM's dominance of the U.S. computer industry during the 1960s and 1970s, created a "product family" of computers in different performance and price classes that all utilized a common operating system and other software.

As Flamm (1988) and others have pointed out, the 360 was not a revolutionary product in terms of its hardware technology (it did not incorporate ICs until 1969). But it was a recognition by one of the leading computer producers of the strategic and constraining role of software within the computer industry and represented a commercially successful solution to this technological bottleneck. The IBM 360 became a dominant design within the mainframe computer industry, and a substantial group of U.S. and foreign

---

[87] According to some accounts (Baum 1981), the Rand group got the programming job only after MIT, IBM, and AT&T had all declined it. IBM, for example, was concerned about how it would employ some 2,000 programmers once the project ended.

firms developed mainframe computers and related products (e.g., data-storage products) that were compatible with the 360 product line.[88]

The introduction of the minicomputer accelerated the segmentation of the computer market and the entry by new firms into competition with the established producers of large systems. The development of the minicomputer was made possible by advances in semiconductor components that reduced the costs of central processing units, as well as lower-cost storage technologies. The Digital Equipment Corporation's PDP-1 minicomputer, introduced in 1960, was one of the first commercial computers to be designed with transistor technology. Kenneth Olsen, the founder of the Digital Equipment Corporation, was an alumnus of the Whirlwind project at MIT. Exploiting a product strategy that reversed that of IBM for the 360, minicomputers were initially sold to sophisticated academic and scientific users who required little software or product support from the manufacturer.

The gradual adoption of the mainframe and minicomputer in industrial applications, such as real-time control of chemicals and petroleum refining processes, contributed to declines in the intensity of energy use per unit of output in these industries (See Schurr et al. 1991, pp. 146–149). Moreover, by supporting more effective modeling and simulation of new processes, computers made possible the smoother introduction of new manufacturing processes into commercial use. The use of "pilot plants" in chemicals and petroleum refining, for example, appears to have declined in importance as a result of better theoretical understanding and real-time

---

[88] The power of software to make or break the commercial success of plug-compatible mainframe computers is illustrated by the experience of RCA, which introduced its Spectra 70 series of computers in 1966. Although they offered comparable performance at lower prices, these machines could not utilize software written for the IBM 360 and ultimately were commercial failures (Flamm 1988).

control. Widespread adoption of computerized real-time control of complex industrial processes, however, required less expensive computers, such as minicomputers, that could be employed in a decentralized computing organization.[89]

The expansion of the overall market for mainframe computers, and (of greater importance) growth in new segments of the computer market (including minicomputers and scientific computers) transformed the structure of the U.S. computer industry. The dominance of the industry by incumbents from the office equipment and related industries faded, and new firms entered. By 1982, just before the onslaught of the desktop computer, four of the ten largest U.S. computer firms were less than fifty years old, and three of these four firms had been founded since 1950 (Table 12). By 1986, new firms accounted for five of the ten largest U.S. computer producers. The rapid growth of the desktop computer market accelerated this transformation and severely undermined the competitive fortunes of four of the five largest producers of computers (IBM, DEC, Unisys, and NCR, which was acquired by AT&T in 1991) in 1986. The seventh-ranking producer in 1986, Wang, was driven into bankruptcy in 1993 by competition from desktop computers. The entry of new firms in this industry, however, typically was driven by the emergence of a new market segment for computer applications. Thus, the dominance of the IBM 360 and 370 was not overturned by direct competition but by the expansion of near-substitutes in the minicomputer and (eventually) desktop computer workstation markets. Rather than the displacement of a "dominant

---

[89] Schurr et al. (1991) note that "large central control computers were very expensive and thus suitable only for large plants ... Moreover, since it was possible for the malfunction of a single component to bring the computer to a halt or cause improper operation, plant managers insisted on having costly backup control systems. By the early 1970s it was apparent that complexity, vulnerability, and resulting high costs had slowed any trend toward centralized computer control in the process industries" (p. 55).

Table 12. *Data Processing Revenues for U.S. Computer Firms, 1963–93.*

	Revenue, *millions of current dollars*				
Firm	1963	1973	1983	1986	1993
IBM	1244	8695	31500	49591	62716
Burroughs	42	1091	3848	—*	—*
Sperry	145	958	2801	9431[†]	7200
Digital	10	265	4019	8414	13637
Hewlett-Packard	—	165	2165	4500	15600
NCR	31	726	3173	4378	9860
Control Data	85	929	3301	3347	452
Scientific Data Systems/Xerox	8	60	—	2100	3330
Honeywell	27	1147	1685	1890	—
Data General	—	53	804	1288	1059
Amdahl	—	—	462	967	1680
General Electric	39	174	862	900	684
Cray Research	—	—	141	597	895
Philco	74	—	—	—	—

* Merged with Sperry.
† Unisys.
Source: Flamm (1988), p. 102; sales data for 1993 are taken from "The Datamation Global 100," *Datamation* 6/15/94, p. 46.

design," this industry has witnessed the fragmentation of markets once dominated by a single design or architecture.

The data in Fig. 13 on trends in the value of shipments of mainframe and minicomputers between 1960 and 1990 depict the rapid increase in minicomputer sales through roughly the mid-1980s, as well as the stagnation in the value of mainframe computer

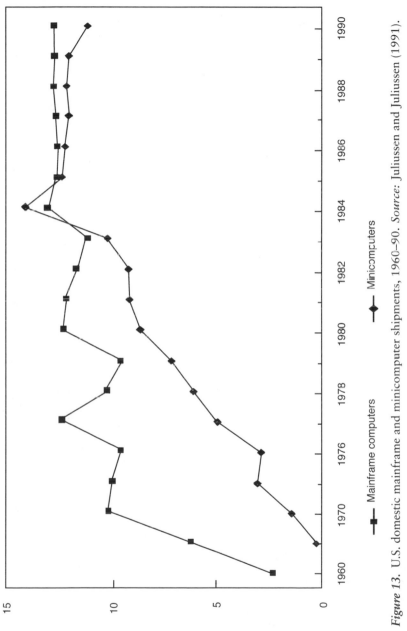

Figure 13. U.S. domestic mainframe and minicomputer shipments, 1960–90. *Source:* Juliussen and Juliussen (1991).

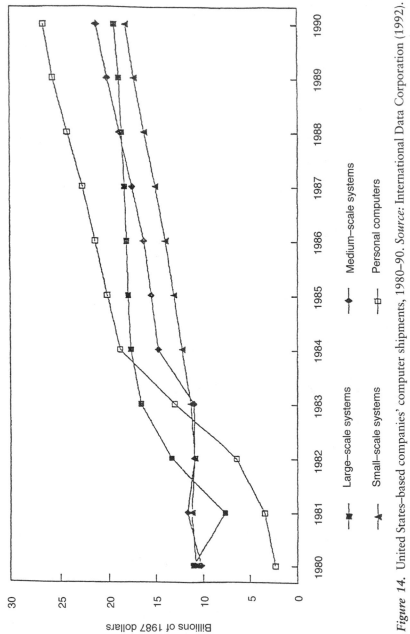

Figure 14. United States–based companies' computer shipments, 1980–90. *Source:* International Data Corporation (1992).

Large–scale systems   Medium–scale systems

Small–scale systems   Personal computers

Billions of 1987 dollars

shipments after the early 1980s. The lack of growth in mainframe sales after the mid-1980s coincided with the rise of the next major segment in this industry, the microcomputer. The rapid increase in microcomputer sales is apparent in Fig. 14, which displays United States–based firms' shipments of "large," "medium," and "small" computers (corresponding to mainframes and larger and smaller minicomputers), as well as personal computers, between 1980 and 1990. This figure depicts even more dramatically than Fig. 12 the decline in rates of growth in shipments of the very largest mainframe computers after 1984. Both figures understate the rate of adoption of computer technology, because they do not adjust the value of shipments for improvements in the power of these computers.

The Microprocessor

The Intel Corporation's commercialization of the integrated circuit microprocessor in 1971 transformed the structure of the U.S. computer industry during the next 25 years. Like the IBM 360, which economized on scarce software development talent, development of the microprocessor at Intel resulted from a search for an integrated circuit that could be used in a wide array of applications. Rather than designing a custom "chipset" for each application, the microprocessor made it possible for Intel to produce a powerful, general-purpose solution to many diverse applications. The microprocessor economized on another scarce resource – engineering design talent that was being squandered on the development of specialized components for each new application (Reid 1984; Slater 1987).[90]

---

[90] Reid (1984, p. 141) quotes the description by Noyce and Hoff (1981) of Intel Corporation of the costs of proliferation in specialized circuit designs and architectures: "If this continued, the number of circuits needed would proliferate beyond the number of circuit designers ... Increased design cost and

The microprocessor broke a bottleneck that limited technological progress and slowed the diffusion of computer technologies.

The microprocessor enabled computing technology to be applied to an unprecedented number and diversity of uses, accelerating its incorporation into the products of such mature manufacturing industries as automobiles, watchmaking, and home appliances, and supported continued growth in applications of computers in such service industries as banking and retailing. In addition, the microprocessor facilitated the development of decentralized computing technologies for control of complex industrial processes in both the materials-processing industries, such as chemicals and petroleum refining, and in mass-production manufacturing (e.g., through the use of microprocessors and programmable logic controllers in machine tools; see Schurr et al. [1991, pp. 298–307]).

By making possible the development of desktop computers and workstations, as well as the development of "massively parallel" supercomputers, the microprocessor supported the entry of new competitors to established producers (in desktop computers, Apple and Compaq challenged IBM; in supercomputers, Intel Corporation, Maspar, and, briefly, Thinking Machines Corporation challenged Cray Computer and Control Data Corporation). The microprocessor enabled firms to amortize the high fixed costs of development and production facilities over longer production runs; its application in the desktop computer had a similar effect on the computer software industry.

diminished usage would prevent manufacturers from amortizing costs over a large user population and would cut off the advantages of the learning curve." For all its potential as a "general purpose technology," however, with great potential for applications in many products, the Intel Corporation's management was slow to recognize the microprocessor's possibilities. Indeed, the firm initially granted an exclusive license for the original microprocessor design to the Japanese electronic calculator firm that commissioned the design (Reid 1984, pp. 140–141).

The Growth of the U.S. Computer Software Industry[91]

The diffusion of microprocessor-based computing technology created huge markets for producers of standardized ("packaged") computer software for desktop computers and workstations. By the 1980s, the rapid and interdependent development of the semiconductor and computer industries had laid the groundwork for the expansion of another "new" postwar industry, the production of standardized computer software for sale in the market (as opposed to its production for internal use). Estimates of the size and recent growth of the U.S. software market are unreliable, because of the poor quality of official statistics and the blurring of the boundaries among "hardware," "software," and "computer services." One recent estimate suggests that in constant (1987) dollars, commercial software revenues in the U.S. market grew from $1.4 billion in 1970 to almost nearly $27 billion in 1988, or nearly twenty-fold (Juliussen and Juliussen 1991).

The growth of the U.S. computer software industry has been marked by at least four distinct eras, the last of which has only begun. During the early years of the first era (1945–65), covering the development and early commercialization of the computer, software as it is currently known did not exist. The concept of computer software as a distinguishable component of a computer system was effectively born with the advent of the von Neumann architecture for stored-program computers. But even after the von Neumann scheme became dominant in the 1950s, software remained closely bound to hardware. The development of a U.S. software industry really began only when computers appeared in significant numbers. The large commercial market for computers that was created by

---

[91] A more detailed discussion of the U.S. and other industrial nations' software industries, on which this section draws, may be found in Mowery (1998).

the IBM 650 provided strong incentives for industry to develop standard software for this architecture.

Along with the development by IBM and other major hardware producers of standard languages such as COBOL and FORTRAN, widespread adoption of a single platform contributed to substantial growth of "internal" software production by large users. But the primary suppliers of the software and services for mainframe computers well into the 1960s were the manufacturers of these machines. In the case of IBM, which leased many of its machines, the costs of software and services were "bundled" with the lease payments. By the late 1950s, however, a number of independent firms had entered the custom software industry. These firms included the Computer Usage Company and Computer Sciences Corporation, both of which were founded by former IBM employees (Campbell-Kelly 1995). Many more independent firms entered the mainframe software industry during the 1960s.

The second era (1965–78) witnessed the first entry of independent software vendors into the industry. During the late 1960s, U.S. producers of mainframe computers began to "unbundle" their software product offerings from their hardware products, separating the pricing and distribution of hardware and software. This development provided opportunities for entry by independent producers of standard and custom operating systems, as well as independent suppliers of applications software for mainframes.

Although independent suppliers of software began to enter the industry in significant numbers in the early 1970s in the United States, computer manufacturers and users remained important sources of both custom and standard software in Japan, Western Europe, and the United States during this period. Some service bureaus that had provided users with operating services and programming solutions began to unbundle their services from their software, providing yet another cohort of entrants into the independent development and sale of traded software. Sophisticated users of computer systems, especially users of mainframe computers, also developed expertise in the creation of solutions to their applications and operating system needs. A number of leading U.S. suppliers of traded software were founded by computer specialists formerly employed by major mainframe users.

Steinmueller (1996) argues that several developments contributed to the emergence of a large independent software industry in the United States during the 1960s. IBM's introduction of the 360 in 1965 provided a single mainframe architecture that utilized a standard operating system spanning all machines in this product family. This development increased the size of the installed base of mainframe computers that could use packaged software designed to operate specific applications and made entry by independent developers more attractive. IBM unbundled its pricing and supply of software and services in 1968, a decision that was encouraged by the threat of antitrust prosecution.[92] The "unbundling" of its software by the dominant manufacturer of hardware (a firm that remains among the leading software suppliers worldwide) provided opportunities for the growth of independent software vendors. Finally, the introduction of the minicomputer in the mid-1960s by firms that typically did not provide bundled software and services opened up another market segment for independent software vendors.

During the third era (1978–93), the development and diffusion of the desktop computer produced explosive growth in the traded software industry. Once again, the United States was the "first mover" in this transformation, and the U.S. market quickly emerged as the largest single one for such packaged software. Rapid adoption

---

[92] As the U.S. International Trade Commission (1995, p. 2-2) pointed out in its recent study, U.S. government procurement of computer services from independent suppliers aided the growth of a sizeable population of such firms by the late 1960s. These firms were among the first entrants into the provision of custom software for mainframe computers after IBM's unbundling of services and software.

of the desktop computer in the United States supported the early emergence of a few dominant designs in desktop computer architecture, creating the first mass market for packaged software. The independent software vendors (ISVs) that entered during this period were largely new to the industry. Few of the major suppliers of desktop software came from the ranks of the leading independent producers of mainframe and minicomputer software, and mainframe and minicomputer ISVs are still minor factors in desktop software.

Both the entry of independent software vendors and the rise to dominance of the IBM PC architecture were linked to IBM's decision to obtain most of the components for its microcomputer from external vendors, including Intel (supplier of the microprocessor) and Microsoft (supplier of the PC operating system, MS-DOS), without forcing them to restrict sales of these components to other producers. The decision to purchase the operating system software from Microsoft was driven by two factors. Development of the IBM PC was a "crash program," undertaken by an autonomous business unit that had insufficient staff or time to undertake in-house development of a family of components or a unique operating system. Equally important, however, was IBM's concern that the PC operate the large number of applications and other programs developed for Microsoft's BASIC operating system. In fact, early IBM PCs contained both the MS-DOS and BASIC operating systems software.[93]

Rapid diffusion of low-cost desktop computer hardware, combined with the rapid emergence of a few dominant designs for this architecture, eroded vertical integration between hardware and software producers and opened up opportunities for ISVs. Reductions in the cost of computing technology have continually expanded the array of potential applications for computers; many of these applications rely on software solutions for their realization.

---

[93] This discussion owes a considerable debt to Professor Thomas Cottrell of the University of Calgary; see Cottrell (1995, 1996).

A growing installed base of ever-cheaper computers has been an important source of dynamism and entry into the traded software industry, because the rapid expansion of market niches in applications has outrun the ability of established computer manufacturers and major producers of packaged software to supply them.[94]

The desktop computer software industry that emerged in the United States had a cost structure that resembled that of the publishing and entertainment industries much more than that of custom software – the returns to a product that was a "hit" were enormous and production costs were relatively low. And like these other industries, the growth of a mass market for software elevated the importance of formal intellectual property rights, especially copyright and patent protection. A major contrast between software and the publishing and entertainment industries, however, is the importance of product standards and consumption externalities in the software market. Users in the mass software market often resist switching among operating systems or even well-established applications because of the high costs of learning new skills, as well as concern over the availability of an abundant library of applications software that complements an operating system. These switching costs, which typically are higher for the less-skilled users who dominate mass markets for software, support the rapid development of "bandwagons" and the creation through market forces of product standards. During the 1980s, these de facto product standards in hardware and software became even more important to the commercial fortunes of software producers than was true during the 1960s and 1970s.

As of 1985, "packaged" software (standard software for use in mainframes, personal, or minicomputers) accounted for more than 75% of the traded software in the U.S. domestic market, and its

---

[94] Bresnahan and Greenstein (1995) point out that a similar erosion of multiproduct economies of scope appears to have occurred among computer hardware manufacturers with the introduction of the microcomputer.

share of domestic consumption has almost certainly grown considerably since that date (Mowery 1998; OECD 1989). Domestic consumption of packaged software has grown rapidly, as desktop computers have diffused widely within the United States. From slightly more than $16 billion in 1985 (in 1992 dollars), the U.S. market for packaged software grew at an average annual rate of slightly more than 10%, to $33.9 billion in 1994 and $46.2 billion in 1996; U.S. Commerce Department data released in late 1997 projected that domestic consumption would exceed $52 billion in 1997 (U.S. Department of Commerce 1997, p. 28-4).[95] Although consumption of packaged software has grown rapidly in other industrial economies, foreign markets remain considerably smaller than that of the United States. Estimated consumption of packaged software in Western Europe in 1996 was $32 billion, and the Japanese packaged software market amounted to only $11.4 billion in that year (U.S. Department of Commerce 1997, p. 28-4).

The large size of the U.S. packaged software market, as well as the fact that it was the first large market to experience rapid growth (reflecting the earlier appearance and rapid diffusion of mainframe and minicomputers, followed by the explosive growth of desktop computer use during the 1980s), gave the U.S. firms that pioneered in

---

[95] Measuring the overall size of the U.S. computer software industry is difficult – its relative youth and limited public statistical agency budgets mean that longitudinal data are very scarce. In addition, the complex structure of the software industry complicates the measurement of industry output, even if one ignores problems of definition and quality adjustment. For example, many firms provide both custom software and computer services, making it difficult to separate the share of output accounted for by software alone. Nevertheless, the available data suggest that the packaged software segment of this industry now is growing more rapidly than other product areas. According to the 1997 *U.S. Statistical Abstract*, "computer programming services," which includes many firms that produce custom software that is developed for specific customers and applications, grew from $22.7 billion in 1990 to $34.8 billion in 1995, a slower rate of growth than packaged software (U.S. Bureau of the Census 1997).

their domestic packaged software market formidable "first-mover" advantages that they exploited internationally. U.S. firms' market shares in their home market exceed 80% in most classes of packaged software, and exceed 65% in non-U.S. markets for all but "applications" software.

The fourth era in the development of the software industry (1992–present) has been dominated by the growth of networking among desktop computers, both within enterprises through local area networks linked to a server and among millions of users through the Internet. Networking has opened opportunities for the emergence of new software market segments (for example, the operating system software that is currently installed in desktop computers may reside on the network or the server) the emergence of new dominant designs,[96] and potentially the erosion of currently dominant software firms' positions. Some network applications that are growing rapidly, such as the World Wide Web, use software (HTML) that operates equally effectively on all platforms, rather than being "locked into" a single architecture. Like the previous eras of this industry's development, the growth of network users and applications has been more rapid in the United States than in

---

[96] In early November 1995, Microsoft, the dominant firm in operating systems software for desktop microcomputers, announced its acquisition of Netwise, a leading supplier of networking software. According to an analyst quoted in a wire-service report on the acquisition, Microsoft's acquisition of Netwise was based on the recognition that "'The desktop paradigm is breaking down and the network is indeed the computer ... The leaders of the stand-alone PC are bowing to the new paradigm,' Googin said. Microsoft is the king of the desktop and the desktop has to be connected" (Reuters News Service 1995). Microsoft has continued to acquire firms and to develop technologies for networking applications. The firm also has entered the development and distribution of content in ventures that seek to exploit the potential convergence of desktop computing and mass entertainment. The firm has entered into an alliance with the NBC television network to form the MSNBC cable channel, and in April 1997, it acquired WebTV, a U.S. firm that has developed technologies that enable television sets to connect to the Internet.

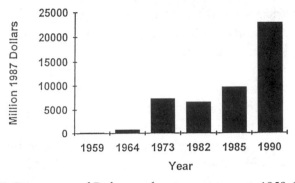

Figure 15. Department of Defense software procurement, 1959–90. *Source:* Langlois and Mowery (1996).

other industrial economies, and U.S. firms have maintained dominant positions in these markets.

As was the case in semiconductors and computer hardware, the U.S. computer software industry sold a large share of its output to federal government agencies, especially the Department of Defense, in its early years. There exists no reliable time series of Department of Defense expenditures on software procurement that employs a consistent definition of software, such as separating embedded software from custom applications or operating systems and packaged software. The data on software expenditures in Fig. 15 are also inconsistent in their treatment of Department of Defense expenditures on software maintenance, as opposed to procurement.

Nevertheless, the trends in these data are dramatic – in constant-dollar terms, Department of Defense expenditures on software appear to have increased more than thirty-fold in just over 25 years, from 1964 through 1990. Throughout this period, Department of Defense software demand was dominated by custom software, and Department of Defense and federal government demand for custom software accounted for a substantial share of the total revenues in this segment of the U.S. software industry. Much of the rapid growth in custom software firms during the period from 1969 through 1980 reflected expansion in federal demand, which in turn was dominated

160

by Department of Defense demand. But like the semiconductor industry, defense markets gradually were outstripped by commercial markets, although the overtaking of defense by commercial demand for software appears to have taken a longer time. By the early 1990s, defense demand accounted for a declining share of the U.S. software industry's revenues.

Its declining share of total demand by the 1990s meant that the defense market no longer exerted sufficient influence on the path of R&D and product development to benefit from generic academic research and product development – defense and commercial needs had diverged. Another illustration of this tendency is the fate of the Defense Department's "generic" software language, Ada, which was unveiled in 1984. Billed as a solution to the severe problems of system maintenance and software development that had resulted from the bewildering variety of software languages in use within defense systems, Ada was intended to serve as a language that could be employed in all defense applications, and one that would attract sufficient interest from commercial developers to produce software that could be used in both civilian and military applications. But the Ada language failed to attract the attention of commercial developers and as a result has languished.

Although demand conditions were favorable, the emergence of a vigorous independent software vendor industry in the United States rested on a research and personnel infrastructure that had benefited from an R&D infrastructure created by federal investments. Perhaps the most important result of these investments was the development of a large university-based research complex that provided a steady stream of new ideas, some new products, and a large number of entrepreneurs and engineers anxious to participate in this industry. Like postwar defense-related funding of R&D and procurement in semiconductors, federal policy toward the software industry was motivated mainly by national security concerns; nevertheless, federal financial support for a broad-based research infrastructure proved quite effective in spawning a vigorous civilian industry.

The emergent U.S. computer software industry was a major beneficiary of postwar federal R&D spending in computer science R&D in universities and industry. Defense-related R&D spending in software appears to have declined somewhat in the 1980s, even as civilian agencies such as the National Science Foundation increased their computer science research budgets. The defense share of federal computer science R&D funding declined from almost 60% in fiscal 1986 to less than 30% in fiscal 1990 (Clement 1987, 1989; Clement and Edgar 1988), and defense funding of computer science R&D in universities in particular appears to have been supplanted somewhat by the growth in funding for quasiacademic research and training organizations.

U.S. antitrust policy also played an important role in this industry's development. The unbundling of software from hardware was almost certainly hastened by the threat of antitrust action against IBM in the late 1960s. Moreover, as is noted in previous discussion, many of the independent vendors who responded to the opportunities created by the new IBM policy had been suppliers of computer services to federal government agencies. The current explosive growth in network applications and Internet-based software and other products has benefited from the restructuring and deregulation of the U.S. telecommunications industry that took place in 1984 as a result of the settlement of the federal antitrust suit against AT&T. The future of the U.S. software industry will also be influenced by the federal government's antitrust oversight of such large software firms as Microsoft. In addition, the relatively liberal U.S. policy toward imports of computer hardware and components supported rapid declines in price-performance ratios in most areas of computer hardware and thereby accelerated domestic adoption of the hardware platforms that provided the mass markets for software producers. Western European and Japanese governments' protection of their regional hardware industries has been associated with higher hardware costs and slower rates of domestic adoption, impeding the growth of their domestic software markets.

A comparison of the U.S. and Western European computer software industries suggests that (especially in standard software) a strong domestic hardware industry is necessary to support the growth of a strong domestic software industry. Comparing the U.S. and Japanese software industries, however, requires that this conclusion be qualified. Japan's computer hardware industry is much stronger than that of Western Europe. But Japanese strength in computer hardware has not been translated into strength in traded software. Japan's strength in computer hardware, which has facilitated the development of competing architectures in mainframes, minicomputers, and microcomputers, in fact appears to have retarded the growth of a domestic packaged software industry. As in Western Europe, Japan's domestic software industry is strongest in the development of custom software solutions (by either hardware manufacturers or independent firms) that rely on close familiarity with user needs.

U.S. software producers derived competitive advantages from their links with the dominant global producers of computer hardware in the early development of mainframe, minicomputer, and desktop systems. The central position of the U.S. market as the "testbed" for developing new applications in such areas as networking and the Internet reflects the enduring importance of user-producer interactions in the software industry. Regardless of the national origin of the hardware on which new software operates, software firms located in the United States are likely to continue to enjoy advantages over firms without a presence in this market.

Conclusion

The postwar development of the U.S. electronics industry (including computer hardware and software, as well as semiconductors) illustrates the broad themes that have characterized U.S. technological development throughout this century, even as it highlights some of the changes in the structure of the innovation process wrought by World War II. Like electricity, the postwar electronics revolution has derived much of its economic impact from a complex and lengthy process of interindustry diffusion and adoption. The products of these high-technology industries have transformed the structure of mature industries (e.g., retailing) as well as newer ones (e.g., commercial aircraft design). The sectoral and economy-wide productivity consequences of computer-based technologies have been realized only gradually, if at all, in large part because of the demanding requirements of these technologies for far-reaching changes in intrafirm organization (David 1990).

Another similarity between the development of electronics and the other technologies discussed in this chapter is the complex and iterative nature of the science-technology interface within each. Fundamental advances in scientific understanding have in many cases been sparked by concerns over the technological performance of such complex systems as long-distance telephony or early warnings of strategic bomber attacks. At the same time, the semiconductor industry in particular has significantly improved the performance and quality of its products, in spite of a very imperfect scientific understanding of the extraordinarily complex physical transformations that underpin its production processes. Pilot plants still play a prominent role in the introduction of new manufacturing processes in semiconductors, reflecting the continuing importance of "cut-and-try" methods of problem solving.

The development of the U.S. electronics industry complex also illustrates a fundamental change in the nature of the U.S. "resource endowment" and its relationship to technological innovation. Expansion in industrial and residential demand for electric power required a more intensive use of fossil fuels and hydroelectric sources for the generation of such power. As we noted in the previous chapter, an important influence on the development of turbines and related technologies for hydroelectric power generation in the United

States was precisely the "discovery" of an abundance of sources for such power.

In contrast, the postwar electronics industry, based as it was on solid-state technologies, did not produce a large surge in demand for natural resources. But the development of the computer software, hardware, and semiconductor industries assuredly did benefit from the abundance of scientific and engineering human capital in the postwar United States, as well as an unusual mix of public and private demand for electronics technologies. The creation of an institutional infrastructure during this century that, by the 1940s, was capable of training large numbers of electrical engineers, physicists, metallurgists, mathematicians, and other experts capable of advancing these new technologies, meant that the postwar American endowment of specialized human capital was initially more abundant than that of other industrial nations. A central factor in this domestic abundance of human capital was the significant increase in the share of the college-educated population that occurred in the aftermath of World War II. In the postwar era, the resource base for knowledge-based industries in electronics, no less than in chemicals, pharmaceuticals, or even automobiles, was transformed. Natural resources per se played a less central role, and the domestic creation of human capital, combined with cross-border flows of knowledge and capital, became indispensable. The postwar United States economy was one of the first illustrations of this trend, which now characterizes much of the global economy.

The development of these sectors also differs in important respects that reveal some significant changes in the structure of the U.S. R&D system. University-based researchers and engineers played an important role in the development of both the U.S. electricity and electronics sectors during this century. But the universities' role in electricity relied primarily on state government and industrial funding, whereas in electronics, federal (mainly defense-related) funding figured very prominently. The federal government also was a major source of early demand for the products of the semiconductor, computer, and software industries, in contrast to its role in the development of electrical technology. A final significant difference between these two sectors concerns the role of new entrants, which were far more significant in the postwar electronics industry than was true of the electrical machinery industry.

7

Concluding Observations

TECHNOLOGICAL CHANGE in the 20th-century United States is best understood in the context of a number of favorable and distinctive initial conditions. Among the most important of these was the rich natural-resource base of the U.S. economy. The direction and impact of technological change within this economy were shaped by the fact that the United States was well endowed by nature with the resources that were essential to modern industrialization.

This kindness of Providence to Americans is well known and has often been commented on, but this characterization is seriously incomplete in one sense. Although one may speak of resources as an endowment provided by nature, one must distinguish between natural resources as a geologist would think of them in surveying a new continent and resources in the much stricter sense of the economist. In 1900 oil that was thousands of feet below the sea floor off the coast of Louisiana would not have constituted a resource to the economist, even if the geologist was aware of its presence, simply because the technology required for its extraction did not yet exist.

The point is that natural resources do not intrinsically possess economic value. That value is a function of the availability of technological knowledge that allows those resources to be extracted and subsequently exploited in the fulfillment of human needs (Rosenberg 1972). As David and Wright (1996) have more recently pointed out, the 19th-century United States was unusual in the speed with which its mineral reserves, the existence of which in many cases had been discovered only a few years earlier, were exploited. The speed with which these reserves were discovered and exploited rested in part on the growth of a substantial university-based apparatus for training mining engineers, anticipating trends observed in the 20th century in the fields of chemical engineering and computer science.

The path of technological innovation in the 19th- and 20th-century United States also contributed to the exploitation of these mineral reserves by creating opportunities for the transformation of lower-quality or valueless raw materials into commercially useful products. Thus, the electric arc furnace converted bauxite from an ore of no economic significance to a valued source of a new metal with commercially attractive characteristics. The same electric arc furnace converted scrapped automobiles into a low-cost source of steel. The Haber-Bosch process converted atmospheric nitrogen into an unlimited source of fertilizer. The automobile and the chemical engineer transformed petroleum from a resource of modest importance as an illuminant to a resource of immense economic significance in transportation, industrial materials, and textiles.

The conclusion, of course, is that natural resources acquire economic value only as a result of the development of technological capabilities that are by no means provided by nature. Twentieth-century American industrial development has consisted in part of learning new techniques for creating and then extracting value from natural resources that had little or no value at the beginning of the century (Rosenberg 1972, chs. 1, 2). In this sense, 20th-century technological change may be characterized as "resource augmenting" in the United States. But the postwar period in particular has also been characterized by far greater reliance on specialized human capital, an input in which the postwar United States was abundantly endowed as the result of investments in a large research and training infrastructure.

Nations lacking domestic natural resources have little choice but to acquire them from foreign sources, something that was difficult for much of this century because of war and economic turmoil. But the creation of a domestic stock of specialized human capital relies on a relatively abundant resource, human intelligence, and energy. As other nations have undertaken similarly large investments in the creation of specialized domestic human capital, and as their access to natural resource imports has improved during the postwar period, the natural resource basis for U.S. comparative advantage has lost much of its significance (Nelson and Wright 1994).

The trajectory of American 20th-century technology was traced along paths that were shaped not only by an abundance of natural resources but also by a large population that was already affluent by contemporary European standards before the First World War and enjoyed a more equal household income distribution. Furthermore, the less-pronounced divisions of social class created a large market for standardized, homogeneous products, as is apparent in the speed with which America came to dominate the world automobile industry. In both autos and commercial aircraft, the geographic dispersion of the U.S. domestic population provided a further impetus to rapid adoption of these technologies.

American domination of automobiles is connected to two other features that shaped the American technological trajectory. The first was the large size of the domestic market within which goods could be freely traded and more complex patterns of industrial specialization could be established. In 1900, the U.S. domestic market was already considerably larger than that of any European country, a fact that increased in economic significance during the period of severe disruption in international flows of goods, capital, and technology from 1914 through 1945. The U.S. domestic market was sufficiently large that American firms, whether in automobiles or chemicals, were better able than European competitors to take maximum advantage of economies of scale during the pre-1945 period. Since World War II, an important factor in improved European and Japanese economic performance has been the revival of world trade, which has reduced the penalties associated with small domestic markets. Nevertheless, both the semiconductor and computer industries derived significant competitive advantage from the large U.S. domestic market during the postwar period.

Another, related feature has recently acquired the name "path dependence," although the phenomenon is as old as the writing of history. The technological competence of a firm or nation at any point in time is shaped by the path that has delivered the economy to its present state. Moreover, this state shapes the ease or difficulty with which different possible future paths of technological development can be exploited. Thus, American success in the 20th-century automobile industry rested in part on the skills of 19th-century U.S. firms in the design and utilization of specialized tools for the manufacture of interchangeable metal components that were assembled into standardized final products.

Other industries, such as the production of boots and shoes, used similar sequences of highly specialized machines, and food processing contributed to the triumph of progressive assembly line technology by demonstrating the feasibility of continuous process "disassembly" of animals at abattoirs and meatpacking plants (Rosenberg 1969, 1972; Hounshell 1982). Ford's assembly line created a technology that was in turn applied to an expanding range of new products in the U.S. economy. The development of "Fordist" assembly-line technology, as well as American leadership in its development, can be understood only as part of a process in which historical sequences mattered a great deal.

This volume has emphasized the importance of interindustry relationships in the analysis of technological change. Although for many purposes the notion of an industry serves as the essential tool of economic analysis, new 20th-century technologies have notoriously failed to respect the sanctity of industrial boundary lines. Intersectoral flows of technology have been around for centuries, but they have become a central feature of 20th-century innovation in the United States. This has been quintessentially the case with the two sectors that have accounted for a rising share of inventive activity in

170

the course of this century as the relative importance of mechanical invention has declined, electricity/electronics and chemicals.

All industries now use electricity to some extent, if only for lighting or to provide power to their computers. There cannot be many sectors of the economy that make no use of at least two chemical intermediate products, paint and plastic. Microprocessors are invading more and more industries, and in 1996 the average new American car contained more than $1,500 worth of electronics components and systems. Indeed, the automobile industry has been a massive importer of new and improved technologies from electronics in recent years, and in earlier decades as well as the present, from machine tools, glass, rubber, paint, plastics, and steel to aluminum and numerous alloys. In providing a large market for such products, the auto industry has strengthened the incentives for supplier firms to pursue innovations. Indeed, the history of industrial research in the United States, especially the distribution of research labs among industries and the character of technological change, would have been dramatically different had the automobile never been invented.

Understanding the impact of technological change involves a close examination of these interindustry technology flows and the reasons for the speed at which they diffuse. In the case of electricity, for example, which became a ubiquitous technology, its early diffusion was very slow for reasons discussed earlier. Although new technologies at any time tend to enter the economy through just a few doors, their entry is followed by widespread diffusion. Consequently, an exclusive focus on the "high-technology" sectors therefore provides an incomplete and distorted view of the economic impacts of new technologies. The textiles industry has long made use of "high-tech" synthetic fibers, purchases of which by the apparel industry now exceed this industry's expenditures on natural fibers. Moreover, the textiles industry now is exploiting lasers and robotics. Although aircraft and forest products are far from one another on the scale of technological intensity, both have benefited extensively from metallurgical improvements as well as from electronic and computer innovations.

In sum, new technologies need to be examined not only at their initial points of entry. A thorough analysis of their histories points out the potential of high technology to revitalize "old" industries, including textiles and forest products, banking and finance, retailing, and medical care. Consider the telephone, invented in the 1870s but adopted widely only in the course of the 20th century. In the post–World War II years, the telephone became ubiquitous in households as well as in businesses – 87% of U.S. households had telephone service by 1970, and by 1994 the number had risen to 94% (U.S. Bureau of the Census 1995). Although the telephone might appear, after more than a century, to be a prime example of a product entering into the mature, slow-growth stage of the life cycle of product innovation, the experience of recent years has been quite different. Despite the fact that the telephone was patented in 1876, it has lately served as a platform on which a variety of more sophisticated communication services have been constructed. The usefulness of the telephone – surely one of the most useful of inventions from its inception – has been powerfully enhanced by facsimile transmission, cellular telephony, electronic mail, data transfer, on-line services, voice mail, conference calls, and "800" numbers. Indeed, a powerful impetus to the rapid growth of computer networking technologies and services in the United States after 1980 was the restructuring and deregulation of telecommunications services that began with the settlement of *U.S. v. AT&T* in 1982.

Moreover, some of the most dramatic improvements in telephone services in the past two decades have been entirely invisible to telephone users, although the ease of direct distance dialing and the improved quality of long distance transmission should be readily apparent to anyone whose memory goes back twenty years or so. The best transatlantic telephone cable in 1966 could carry only 138 conversations between Europe and North America simultaneously. The first fiber-optic cables, using lasers for transmission, were installed in 1988 and had the capacity to carry 40,000 conversations simultaneously. The fiber-optic cables installed in the early 1990s have an even greater capacity (Wriston 1992, pp. 43–44).

A large part of the story of technological change in 20th-century America concerns the changing economic role of science. But even this sweeping statement is insufficiently broad. A more accurate and comprehensive formulation would recognize the growing economic importance not just of science but of the broader institutionalization of research. This is certainly more consistent with present-day reality, in which most of what is referred to as "R&D" is something other than science. Roughly two thirds of U.S. R&D investment constitutes "D," which is to say that most R&D expenditures are devoted to product design and testing, redesign, improvements in manufacturing processes, and so forth. Most R&D has not been science, whether basic or applied. Rather, as Whitehead long ago insisted in the quotation in the opening paragraph of our first chapter, most of it represents a search for ways of "bridging the gap between the scientific ideas, and the ultimate product. It is a process of disciplined attack upon one difficulty after another." Throughout the 20th century the United States has enjoyed considerable success in institutionalizing this process of "disciplined attack" within the private industrial firm.

Most discussions of the growing economic role of science in this century have dealt with this growth as if it were a purely exogenous phenomenon – it is assumed that the corpus of science grew for reasons that were independent of economic forces but, once generated, that knowledge was subsequently applied to the solution of economic problems. There is no doubt some truth to this view, but it is very incomplete. There appears to be a deeper and neglected question, which may be referred to as the determinants of the demand for science. Why did the economic "payoff" to the findings of science suddenly increase sufficiently to repay firms' investments in R&D? The analysis of this volume suggests that scientific advance has become less and less a matter of the independent unfolding of knowledge and more and more a response to techno-logical progress in the development of practical means to produce goods and services.

We have already suggested that the American antitrust laws dis-couraged firms from cartel-like forms of rent-seeking behavior of the sort that played a large role in Europe, leading American firms to focus their strategic behavior more strongly on internal activities such as R&D. A further possibility is suggested by the observation that the rapid expansion in the number of industrial research labo-ratories occurred at about the same time – the first three decades of the 20th century – as the growth of university research that was of increasing value to private industry. Thus there was a strong com-plementarity between university research and industrial research, in the sense that the growth of university research raised the ex-pected rate of return to the establishment of an industrial research capability, and vice versa. Here again it is essential to define re-search to include work in the engineering disciplines as well as scientific research of a very applied nature – assaying of mineral ores, determining the strength of building materials or the opti-mal design of an aircraft given its power plant capability, and so forth.

The need to improve the performance of an expanding techno-logical system has shaped and mobilized the research agenda in industrial laboratories to an increasing degree as a result of the in-stitutionalization of organized research within the private industrial firm. A primary mission of these labs has been to exploit scientific knowledge and methodology to reduce costs, to increase product reliability, and to develop entirely new products or manufacturing processes. The problems encountered by sophisticated industrial technologies and the anomalous observations or unexpected diffi-culties they produced have served as powerful stimuli to scientific research in the academic community as well as in the industrial re-search lab. In these ways the responsiveness of scientific research to economic needs and opportunities has been powerfully reinforced.

Thus, solid-state physics, presently the largest subdiscipline of physics, attracted only a few physicists before the advent of the transistor in December 1947, although this small number included some of the most distinguished minds in the profession. The transistor demonstrated the potentially high payoff of solid-state research and led to a huge concentration of resources in that field. The rapid mobilization of intellectual resources in solid-state research after the invention of the transistor occurred in the university as well as in private industry. As we note in a previous chapter, transistor technology did not build on a preexisting academic research commitment of major proportions. But this technological breakthrough led to a large-scale commitment of academic scientific resources to basic research in this field. An analogous case could be made concerning the growing commitment of resources to the subdiscipline of surface physics.

Similarly, the development of the laser, and the possibility of combining the laser with optical-fiber light guides for transmission purposes, pointed forcefully toward optics as a field in which advances in knowledge might be expected to have high payoffs. As a result, optics as a field of scientific research experienced a huge resurgence in the 1960s, immediately following the first successful construction of a ruby laser by the physicist Theodore H. Maiman in 1960. Optics was quickly converted by changed expectations from a relatively quiet intellectual backwater of science to a burgeoning field of research. In this sense, technology has come to influence science in the most powerful of ways: by determining its research agenda.

We noted in our introductory chapter that the technological development of the U.S. economy during this century, no less than during the previous century, must be viewed in its international context. The "globalization" of technology flows that is widely cited as a hallmark of the last quarter of this century has in fact been an important influence on U.S. economic development for well over a century. For much of the pre-1940 period, we have argued, the

United States was an adept technological "borrower" and commercializer, benefiting in many instances from the large size of its domestic market, relatively high and evenly distributed household incomes, and a geographically dispersed population. The scientific and many of the technological breakthroughs underpinning such critically important technological advances as the internal combustion engine, polymer and organic chemicals, and the commercial jet aircraft can all be traced to European sources during the late 19th and 20th centuries. But in many cases U.S. firms were among the most successful commercializers of the products based on these fundamental advances.

Moreover, the successful borrowing and application by U.S. firms of inventions developed elsewhere relied on strength in engineering and technology development, rather than excellence in scientific research. For most of the pre-1940 period, U.S. basic research was of distinctly secondary quality, by comparison with that of such European nations as Germany, the United Kingdom, and France, although American research in physics was clearly becoming world class in the interwar years. The fundamental transformation in the structure of the U.S. R&D system wrought by World War II changed the status of U.S. science from follower to undisputed leader.

If the inventive capabilities of U.S. firms before 1940 did not rest on science, whence did they spring? Innovation relied on U.S. strengths in technology development, manufacturing, marketing, and engineering, largely located in private firms. Moreover, these strengths were themselves developed because basic characteristics of the American economy and society led this nation to a more complete exploitation of the opportunities inherent in a technological path that was relatively resource-intensive and capital-using, but at the same time more scale-dependent than was attainable by European nations. The "resource-intensive" character of much of the innovative activity of U.S. firms throughout the pre-1940 period has been noted here; the U.S. resource base during this period was also augmented by growing public investments in public institutions of higher education, whose research and training activities supported much of the inward technology transfer that underpinned U.S. inventive prowess. The state universities, in particular, were strongly oriented toward the exploitation of the local resource base (Rosenberg and Nelson 1994; David and Wright 1996).

The creation of a large scientific research complex during and after World War II changed the position of the United States within the international R&D system. No longer primarily borrowers or imitators, U.S. firms, drawing on an infrastructure in industry and universities that was financed in large part by federal funds, now became leaders in the invention and early-stage commercialization of new technologies. The structure of such postwar U.S. high-technology industries as microelectronics and computers, characterized by high levels of entry by new as well as established firms, strong competition, and relatively weak protection of firm-specific intellectual property, were well-suited to the task of sorting out the numerous technological alternatives and uncertainties over commercialization posed by these new opportunities.

But the R&D system created by World War II arguably provided no more support to the types of technology development and commercialization activities in which U.S. firms had excelled before the war, and other developments contributed to much more intense competition from foreign sources. In the nature of the case, the basic research investments of the federal government and U.S. firms yielded important advances that (with sufficient investment and skill) could be exploited by non-U.S. firms. Moreover, the commercial returns to the large defense-related investments of the federal government appear to have declined over the course of the postwar period.

The advantages that U.S. firms derived from their large domestic market and access to natural resources also declined, as a result of the revival of the global trading system, a key objective of U.S. foreign economic policy during this period. Improvements in the technologies of travel and communication accelerated international

177

transmission of advances in both technology and science. As a result, the ability of U.S. firms to reap the economic benefits of U.S. leadership in basic science and engineering was weakened during the latter half of the postwar period.

Was 20th-century technological change in this nation largely determined by a unique "American national system of innovation," which differed significantly from those of other industrial nations? Or is this historical pattern instead the result of a set of conditions, elaborated here (including the large domestic market, relative resource abundance, a relatively egalitarian income distribution), that favored a trajectory of economic and technological development that proved to be especially fruitful during this century? In industries such as chemicals, for example, the U.S. resource endowment gave U.S. firms a "head start" during the interwar period in developing the technological and other skills necessary to exploit the possibilities of petroleum-based feedstocks and polymer chemistry for an unprecedented abundance of new products and low-cost manufacturing processes in the postwar era. To cite only one example from the 1980s, the large domestic U.S. market and the dominance of the English language within this and other major markets have provided important advantages to U.S. entrepreneurs and innovators in the computer software industry.

At the same time, however, we noted previously that throughout this century, and especially since World War II, the institutional structure of the U.S. "innovation system" has differed significantly from those of most other industrial economies. But our discussion also has highlighted the fact that the structure of this "American system" has hardly been constant; indeed, the organization of R&D in this economy arguably has undergone more far-reaching structural change during the 20th century than is true of other industrial, capitalist economies.

We therefore conclude by suggesting that both of these broad sets of factors were indispensable in defining an unusual, and unusually fruitful, trajectory of economic and technological development for the United States during the 20th century. Their influences cannot and should not be separated. The unusual institutional structure of the U.S. R&D system, such as the important role of universities in supporting the development of mining engineering, chemical engineering, and petroleum engineering, contributed to the discovery and exploitation of this nation's natural-resource endowment. This conclusion may also provide some basis for guarded optimism about future developments. Although this economy's natural resource endowment no longer defines an important source of comparative advantage, other characteristics of the late–20th century United States, such as its large domestic market, continue to provide competitive advantages in specific products and technologies. We are less certain that the institutional complements to these "natural" advantages will be sustained. But this institutional structure has proven to be highly adaptive. A clearer understanding of its contributions to innovation and economic growth is indispensable to the process of structural change that lies ahead.

Bibliography

Abernathy, W.J., 1978. *The Productivity Dilemma*. (Baltimore, MD: Johns Hopkins University Press).

Abramovitz, M., 1956. "Resource and Output Trends in the United States Since 1870." *American Economic Review* 46, 5–23.

Abramovitz, M., 1986. "Catching Up, Forging Ahead, and Falling Behind." *Journal of Economic History* 46, 385–406.

Abramovitz, M., 1990. "The Catch-Up Factor in Postwar Economic Growth." *Economic Inquiry* 28, 1–18.

Achilladelis, B., A. Schwarzkopf, and M. Cines, 1987. "A Study of Innovation in the Pesticide Industry: Analysis of the Innovation Record of an Industrial Sector." *Research Policy* 16, 175–212.

Aftalion, F., 1991. *A History of the International Chemical Industry*, O. Benfey. trans. (Philadelphia: University of Pennsylvania Press).

Ames, J., 1925. *Statement of NACA Chairman to the President's Aircraft Board*. (Washington, DC: U.S. Government Printing Office), p. 731.

American Chemical Society, 1973. *Chemistry in the Economy*. (Washington, DC: American Chemical Society).

American Institute of Chemical Engineers, 1970. *The History of Penicillin Production*. (New York: American Institute of Chemical Engineers).

Arora, A., and A. Gambardella, 1994. "The Changing Technology of Technological Change." *Research Policy* 23, 523–532.

Arora, A., and N. Rosenberg, 1998. "Chemicals: A U.S. Success Story," in A. Arora, R. Landau, and N. Rosenberg, eds., *Chemicals and Long-Term Economic Growth*. (New York: John Wiley).

Arrow, K.J., 1976. "Classificatory Notes on the Production and Transmission of Technical Knowledge," in *Essays in the Theory of Risk-Bearing*. (New York: American Elsevier).

Bibliography

Aspray, W., and B.O. Williams, 1994. "Arming American Scientists: The Role of the National Science Foundation in the Provision of Scientific Computing Facilities for Colleges and Universities." *Annals of the History of Computing* 16, 60–74.

Baily, M.N., and A.K. Chakrabarti, 1988. *Innovation and the Productivity Crisis.* (Washington, DC: Brookings Institution).

Barfield, C.E., 1982. *Science Policy from Ford to Reagan.* (Washington, DC: American Enterprise Institute).

Barnett, D., and R. Crandall, 1986. *Up From the Ashes: The Rise of the Steel Minimill in the U.S.* (Washington, DC: Brookings Institution).

Baum, C., 1981. *The System Builders: The Story of SDC.* (Santa Monica, CA: System Development Corp.).

Beer, J.H., 1959. *The Emergence of the German Dye Industry.* (Urbana, IL: University of Illinois Press).

Birr, K., 1966. "Science in American Industry," in D. Van Tassel and M. Hall, eds., *Science and Society in the U.S.* (Homewood, IL: Dorsey).

Blumenthal, D., M. Gluck, K.S. Louis, and D. Wise, 1986. "Industrial Support of University Research in Biotechnology." *Science* 231, 242–246.

Bollinger, L., K. Hope, and J.M. Utterback, 1983. "A Review of Literature and Hypotheses on New Technology-Based Firms." *Research Policy* 12, 1–14.

Bond, E.C., and S. Glynn, 1995. "Recent Trends in Support for Biomedical Research and Development," in N. Rosenberg, A.C. Gelijns, and H. Dawkins, eds., *Sources of Medical Technology: Universities and Industry.* (Washington, DC: National Academy Press).

Borrus, M.G., 1988. *Competing for Control.* (Cambridge, MA: Ballinger).

Braun, E., and S. MacDonald, 1982. *Revolution in Miniature*, 2d ed. (New York: Cambridge University Press).

Bresnahan, T.F., and S. Greenstein, 1995. "Technological Competition and the Structure of the Computer Industry." Unpublished manuscript.

Bresnahan, T.F., and M. Trajtenberg, 1995. "General Purpose Technologies: Engines of Growth?" *Journal of Econometrics* 65, 232–244.

Briggs, A., 1981. "Social History 1900–1945," in R. Floud and D.W. McCloskey, eds., *The Economic History of Britain since 1700*, vol. 2. (Cambridge: Cambridge University Press).

Bright, A.A., 1949. *The Electric-Lamp Industry.* (New York: Macmillan).

Brooks, J., 1976. *Telephone: The First Hundred Years.* (New York: Harper and Row).

Burns, A.F., 1934. *Production Trends in the U.S. Since 1870.* (New York: National Bureau of Economic Research).

Bibliography

Bush, V., 1945. *Science: The Endless Frontier*. (Washington, DC: U.S. Government Printing Office).

Business Week, June 26, 1989. "Is the U.S. Selling Its High-Tech Soul to Japan?" 117–118.

Business Week, August 7, 1989. "Advanced Bio Class? That's Over in Hitachi Hall." 73–74.

Campbell-Kelly, M., 1995. "Development and Structure of the International Software Industry, 1950–1990." *Business and Economic History* 24, 73–110.

Carr, C., 1952. *Alcoa*. (New York: Rinehart).

Chandler, A.D., Jr., 1962. *Strategy and Structure: Chapters in the History of Industrial Enterprise*. (Cambridge, MA: MIT Press).

Chandler, A.D., Jr., 1974. "Structure and Investment Decisions in the United States," in H. Daems and H. v. d. Wee, eds., *The Rise of Managerial Capitalism*. (The Hague: Martinus Nijhoff).

Chandler, A.D., Jr., 1976. "The Development of Modern Management Structure in the US and UK," in L. Hannah, ed., *Management Strategy and Business Development*. (London: Macmillan).

Chandler, A.D., Jr., 1977. *The Visible Hand*. (Cambridge, MA: Harvard University Press).

Chandler, A.D., Jr., 1978. "The United States: Evolution of Enterprise," in *The Cambridge Economic History of Europe*, vol. 7, P. Mathias and M. Postan, eds., *The Industrial Economies: Capital, Labour, and Enterprise*, part II. (Cambridge: Cambridge University Press).

Chandler, A.D., Jr., 1980a. "The United States: Seedbed of Managerial Capitalism," in A.D. Chandler and H. Daems, eds., *Managerial Hierarchies*. (Cambridge: Harvard University Press).

Chandler, A.D., Jr., 1980b. "The Growth of the Transnational Industrial Firm in the United States and the United Kingdom: A Comparative Analysis." *Economic History Review* 2d ser., 33, 396–410.

Chandler, A.D., Jr., 1990. *Scale and Scope*. (Cambridge: Harvard University Press).

Chandler, A.D., Jr., T. Hikino, and D.C. Mowery, 1998. "The Development of Corporate Strategy in the World's Largest Chemicals Firms," in A. Arora, R. Landau, and N. Rosenberg, eds., *Chemicals and Long-Term Economic Growth*. (New York: John Wiley).

Chesnais, F., 1988. "Technical Co-Operation Agreements Between Firms," *STI Review* 4, 51–119.

Clark, K., and T. Fujimoto, 1991. *Product Development Performance*. (Boston: Harvard Business School Press).

Bibliography

Clark, V., 1929. *History of Manufactures in the United States*, vol. III, *1893–1928*. (New York: McGraw-Hill).

Clement, J.R.B., 1987. "Computer Science and Engineering Support in the FY 1988 Budget," in Intersociety Working Group, ed., *AAAS Report XII: Research and Development, FY 1988*. (Washington, DC: American Association for the Advancement of Science).

Clement, J.R.B., 1989. "Computer Science and Engineering Support in the FY 1990 Budget," in Intersociety Working Group, ed., *AAAS Report XIV: Research and Development, FY 1990*. (Washington, DC: American Association for the Advancement of Science).

Clement, J.R.B., and D. Edgar, 1988. "Computer Science and Engineering Support in the FY 1989 Budget," in Intersociety Working Group, ed., *AAAS Report XIII: Research and Development, FY 1989*. (Washington, DC: American Association for the Advancement of Science).

Cohen, I.B., 1976. "Science and the Growth of the American Republic." *Review of Politics* 38, 359–398.

Collins, N.R., and L.E. Preston, 1961. "The Size Structure of the Largest Industrial Firms." *American Economic Review* 51, 986–1011.

Committee on Industry and Trade, 1928. *Survey of Metal Industries*, pt. 4. (London: H.M. Stationery Office).

Congressional Budget Office, 1984. *Federal Support for R&D and Innovation*. (Washington, DC: Congressional Budget Office).

Cottrell, T., 1995. "Strategy and Survival in the Microcomputer Software Industry, 1981–1986," unpublished Ph.D. dissertation, Haas School of Business, University of California, Berkeley.

Cottrell, T., 1996. "Standards and the Arrested Development of Japan's Microcomputer Software Industry," in D.C. Mowery, ed., *The International Computer Software Industry: A Comparative Study of Industry Evolution and Structure*. (New York: Oxford University Press).

Danhof, C., 1968. *Government Contracting and Technological Change*. (Washington, DC: The Brookings Institution).

David, P.A., 1975. *Technical Choice, Innovation, and Economic Growth*. (New York: Cambridge University Press).

David, P.A., 1986. "Technology Diffusion, Public Policy, and Industrial Competitiveness," in R. Landau and N. Rosenberg, eds., *The Positive Sum Strategy: Harnessing Technology for Economic Growth*. (Washington, DC: National Academy Press).

David, P.A., 1990. "The Computer and the Dynamo." *American Economic Review* 90, 355–361.

Bibliography

David, P.A., and G. Wright, 1996. "Increasing Returns and the Genesis of American Resource Abundance." CEPR Working Paper #472, Stanford University.

Davis, B., January 4, 1989. "Pentagon Seeks to Spur U.S. Effort to Develop 'High-Definition' TV." *Wall Street Journal*, 29.

Davis, L.E., R.A. Easterlin, and W.N. Parker, eds., 1972. *American Economic Growth: An Economist's History of the United States.* (New York: Harper & Row).

Davis, L.E., and D.C. North, 1971. *Institutional Change and American Economic Growth.* (New York: Cambridge University Press).

Derian, J.-C., 1990. *America's Struggle for Leadership in Technology.* (Cambridge, MA: MIT Press).

Dertouzos, M., R. Lester, and R. Solow, eds., 1989. *Made in America.* (Cambridge: MIT Press).

Du Boff, R., December 1967. "The Introduction of Electric Power in American Manufacturing." *Economic History Review.* ser. 2, 20, 509–518.

Economist, February 10, 1989. "Test-Tube Trauma." 67.

Economist, June 24, 1989. "Venture-Capital Drought." 73–74.

Economist, February 3, 1990. "Out of the Ivory Tower." 65–72.

Edison Electric Institute, various years. *Statistical Yearbook of the Electric Power Utilities Industry.* (New York: Edison Electric Institute).

Edwards, R.C., 1975. "Stages in Corporate Stability and Risks of Corporate Failure." *Journal of Economic History* 35, 418–457.

Encyclopaedia Britannica, 13th ed., supplement.

Encyclopaedia Britannica, 15th ed. "Nobel Prize Winners," 8, 740–747.

Enos, J., 1958. "A Measure of the Rate of Technological Progress in the Petroleum Refining Industry." *Journal of Industrial Economics* 6, 180–197.

Enos, J., 1962. *Petroleum Progress and Profits.* (Cambridge, MA: MIT Press).

Ergas, H., 1987. "Does Technology Policy Matter?" in H. Brooks and B. Guile, eds., *Technology and Global Industry.* (Washington, DC: National Academy Press).

Ernst and Young, Inc., 1996. *Biotech 96: Pursuing Sustainability.* (Palo Alto, CA: Ernst and Young).

Evenson, R.E., 1982. "Agriculture," in R.R. Nelson, ed., *Government and Technical Progress.* (New York: Pergamon).

Evenson, R.E., 1983. "Intellectual Property Rights and Agribusiness Research and Development: Implications for the Public Agricultural Research System." *American Journal of Agricultural Economics* 65, 967–976.

Feldman, M., and Y. Schreuder, 1996. "Initial Advantage: The Origins of the Geographic Concentration of the Pharmaceutical Industry in the Mid-Atlantic Region." *Industrial and Corporate Change* 5, 839–862.

Ferguson, C.H., 1983. "The Microelectronics Industry in Distress." *Technology Review* 86, 24–37.

Ferguson, C.H., 1988. "From the People Who Brought You Voodoo Economics." *Harvard Business Review* 66, 55–62.

Fisher, F.M., J.W. McKie, and R.B. Mancke, 1983. *IBM and The U.S. Data Processing Industry.* (New York: Praeger).

Fishlow, A., 1972. "Internal Transportation," in L. Davis *et al.*, eds., *American Economic Growth: An Economist's History of the United States.* (New York: Harper and Row).

Flamm, K., 1988. *Creating the Computer.* (Washington, DC: Brookings Institution).

Flamm, K., and T. McNaugher, 1989. "Rationalizing Technology Investments," in J.D. Steinbruner, ed., *Restructuring American Foreign Policy.* (Washington, DC: Brookings Institution).

Fligstein, N., 1990. *The Transformation of Corporate Control.* (Cambridge, MA: Harvard University Press).

Flink, J.R., 1970. *America Adopts the Automobile, 1895–1910.* (Cambridge, MA: MIT Press).

Florida, R.L., and M. Kenney, 1988. "Venture Capital-Financed Innovation and Technological Change in the USA." *Research Policy* 17, 119–137.

Florida, R.L., and D.F. Smith, 1990. "Venture Capital, Innovation, and Economic Development." *Economic Development Quarterly* 4, 345–360.

Galambos, L., 1966. *Competition and Cooperation.* (Baltimore, MD: Johns Hopkins University Press).

Galler, B.A., 1986. "The IBM 650 and the Universities." *Annals of the History of Computing* 8, 36–38.

Gambardella, A., 1995. *Science and Innovation.* (New York: Cambridge University Press).

Gelijns, A., and N. Rosenberg, 1998. "Diagnostic Devices: An Analysis of Comparative Advantage," in D.C. Mowery and R.R. Nelson, eds., *The Sources of Industrial Leadership.* (New York: Cambridge University Press).

Gerschenkron, A., 1962. "Economic Backwardness in Historical Perspective," in *Economic Backwardness in Historical Perspective.* (Cambridge: Harvard University Press).

Bibliography

Gibb, G.S., and E.H. Knowlton, 1956. *The Resurgent Years: History of Standard Oil Company (New Jersey), 1911–1927.* (New York: Harper).

Giedion, S., 1948. *Mechanization Takes Command.* (New York: Oxford University Press).

Ginsberg, E., and A.B. Dutka, 1989. *The Financing of Biomedical Research.* (Baltimore, MD: Johns Hopkins University Press).

Goldstine, H., 1972. *The Computer from Pascal to Von Neumann.* (Princeton, NJ: Princeton University Press).

Gomory, R.E., 1988. "Reduction to Practice: The Development and Manufacturing Cycle," in *Industrial R&D and U.S. Technological Leadership.* (Washington, DC: National Academy Press).

Gordon, R.J., 1993. "Forward into the Past: Productivity Regression in Electric Power Generation." NBER Working Paper #3988.

Gorte, J. F., July 13, 1989. Testimony before the Subcommittee on Science, Research, and Technology, Committee on Science, Space, and Technology, U.S. House of Representatives.

Graham, M.B.W., 1986a. *RCA and the Videodisc: The Business of Research.* (Cambridge: Cambridge University Press).

Graham, M.B.W., 1986b. "Corporate Research and Development: The Latest Transformation." *Technology in Society* 7, 179–195.

Graham, M.B.W., 1988. "R&D and Competition in England and the United States: The Case of the Aluminum Dirigible." *Business History Review* 62, 261–285.

Graham, M.B.W., and B.H. Pruitt, 1990. *R&D for Industry: A Century of Technical Innovation at Alcoa.* (New York: Cambridge University Press).

Griliches, Z., 1958. "The Demand for Fertilizer: An Economic Interpretation of a Technical Change." *Journal of Farm Economics* 40, 591–606.

Gruber, C., 1995. "The Overhead System in Government Sponsored Academic Science: Origins and Early Development." *Historical Studies in the Physical and Biological Sciences* 25, 241–268.

Grunwald, J., and K. Flamm, 1985. *The Global Factory.* (Washington, DC: Brookings Institution).

Gupta, U., July 19, 1982. "Biotech Start-Ups Are Increasingly Bred Just to Be Sold." *Wall Street Journal* B2.

Gupta, U., November 20, 1988. "Start-Ups Face Big-Time Legal Artillery." *Wall Street Journal* B2.

Haber, F., 1971. *The Chemical Industry, 1900–1930.* (New York: Oxford University Press).

Hall, B.H., 1988. "The Effect of Takeover Activity on Corporate Research and Development," in A. Auerbach, ed., *Corporate Takeovers: Causes and Consequences.* (Chicago: University of Chicago Press).

Hanle, P., 1982. *Bringing Aerodynamics to America.* (Cambridge, MA: MIT Press).

Harris, R.G., and D.C. Mowery, September/October 1990. "New Plans for Joint Ventures: The Results May Be an Unwelcome Surprise." *The American Enterprise* 1, #5, Sept.–Oct., 52–55.

Hausman, J., 1997. "Cellular Telephones, New Products, and the CPI." NBER Working Paper #5982.

Hawley, E., 1966. *The New Deal and the Problem of Monopoly.* (Princeton, NJ: Princeton University Press).

Hayami, Y., and V. Ruttan, 1971. *Agricultural Development.* (Baltimore: Johns Hopkins Press).

Haynes, W., 1945. *American Chemical Industry*, vol. II, *The World War I Period: 1912–22.* (New York: Van Nostrand).

Heffernan, V., 1997. personal communication.

Heilbron, J.L., and R.W. Seidel, 1989. *Lawrence and His Laboratory: A History of the Lawrence Berkeley Laboratory*, vol. 1. (Berkeley, CA: University of California Press).

Herbert, V., and A. Bisio, 1985. *Synthetic Rubber.* (Westport, CT: Greenwood Press).

Henderson, R., L. Orsenigo, and G. Pisano, 1998. "The Pharmaceutical Industry and the Revolution in Molecular Biology," in D.C. Mowery and R.R. Nelson, eds., *The Sources of Industrial Leadership.* (New York: Cambridge University Press).

Hendry, J., 1989. *Innovating for Failure.* (Cambridge: MIT Press).

Hirsh, R.F., 1989. *Technology and Transformation in the American Electric Utility Industry.* (New York: Cambridge University Press).

Hobby, G.L., 1985. *Penicillin.* (New Haven, CT: Yale University Press).

Hodder, J.E., 1988. "Corporate Capital Structure in the United States and Japan: Financial Intermediation and Implications of Financial Deregulation," in J.B. Shoven, ed., *Government Policy Towards Industry in the United States and Japan.* (New York: Cambridge University Press).

Holley, I.B., 1964. *Buying Aircraft: Materiel Procurement for the Army Air Forces*, vol. 7 of the *Special Studies of the U.S. Army in World War II.* (Washington, DC: U.S. Government Printing Office).

Hounshell, D.A., 1982. *From the American System of Manufacture to Mass Production.* (Baltimore, MD: Johns Hopkins Press).

Bibliography

Hounshell, D.A., and J.K. Smith, October 7, 1985. "Du Pont: Better Things for Better Living Through Research," presented at "The R&D Pioneers," Hagley Museum and Library, Wilmington, DE.

Hounshell, D.A., and J.K. Smith, Jr., 1988. *Science and Corporate Strategy: Du Pont R&D, 1902–1980*. (New York: Cambridge University Press).

Hounshell, D.A., 1992. "Du Pont and the Management of Large-Scale Research and Development," in P. Gallison and B. Hevly, *Big Science: The Growth of Large-Scale Research*. (Stanford, CA: Stanford University Press).

Hounshell, D.A., 1996. "The Evolution of Industrial Research in the United States," in R. Rosenbloom and W.J. Spencer, eds., *Engines of Innovation: U.S. Industrial Research at the End of an Era*. (Boston, MA: Harvard Business School Press).

Hughes, W., 1971. "Scale Frontiers in Electric Power," in William Capron, ed., *Technological Change in Regulated Industries*. (Washington, DC: Brookings Institution).

Hughes, T.P., 1983. *Networks of Power*. (Baltimore, MD: Johns Hopkins University Press).

International Data Corporation, 1992. *Computer Industry Reports: The Gray Report*. (Framingham, MA: International Data Corporation).

Jorde, T.M., and D.J. Teece, 1989. "Competition and Cooperation: Striking the Right Balance." *California Management Review* 31, 25–37.

Joskow, P., 1987. "Productivity Growth and Technical Change in the Generation of Electricity." *Energy Journal* 8, 17–38.

Juliussen, K., and E. Juliussen, 1991. *The Computer Industry Almanac: 1991*. (New York: Simon and Schuster).

Kaplan, A.D.H., 1964. *Big Business in a Competitive System*. (Washington, DC: Brookings Institution).

Katz, B., and A. Phillips, 1982. "The Computer Industry," in Richard R. Nelson, ed., *Government and Technical Progress: A Cross-Industry Analysis*. (New York: Pergamon Press).

Katz, M.L., and J.A. Ordover, 1990. "R&D Competition and Cooperation." *Brookings Papers on Economic Activity: Microeconomics*, 137–192.

Kendrick, J., 1961. *Productivity Trends in the United States*. (Princeton, NJ: Princeton University Press).

Kleiman, H., 1966. *The Integrated Circuit Industry: A Case Study of Product Innovation in the Electronics Industry*, unpublished D.B.A. dissertation, The George Washington University.

Bibliography

Kuznets, S., 1930. *Secular Movements in Production and Prices.* (Boston: Houghton Mifflin).

Kuznets, S., 1959. *Six Lectures on Economic Growth.* (Glencoe, NY: The Free Press).

Lamoreaux, N., 1985. *The Great Merger Movement in American Business, 1895–1904.* (New York: Cambridge University Press).

Landau, R., and N. Rosenberg, 1992. "Successful Commercialization in the Chemical Process Industries," in N. Rosenberg, R. Landau, and D. Mowery, eds., *Technology and the Wealth of Nations.* (Stanford, CA: Stanford University Press).

Landsberg, H., and S. Schurr, 1968. *Energy in the United States.* (New York: Random House).

Langlois, R.N., and D.C. Mowery, 1996. "The Federal Government Role in the Development of the U.S. Software Industry," in D.C. Mowery, ed., *The International Computer Software Industry.* (New York: Oxford University Press).

Larson, H.M., E.H. Knowlton, and C.S. Popple, 1971. *History of Standard Oil (New Jersey): New Horizons, 1927–1950.* (New York: Harper & Row).

Leslie, S., 1993. *The Cold War and American Science.* (New York: Columbia University Press).

Levin, R.C., 1982. "The Semiconductor Industry," in R.R. Nelson, ed., *Government and Technical Progress: A Cross-Industry Comparison.* (New York: Pergamon Press).

Levin, R.C., W.M. Cohen, and D.C. Mowery, 1985. "R&D, Appropriability, Opportunity, and Market Structure: New Evidence on Some Schumpeterian Hypotheses." *American Economic Review Papers and Proceedings* 75, 20–24.

Lewis, W., 1953. "Chemical Engineering: A New Science?" in L.R. Lohr, ed., *Centennial of Engineering: 1852–1952.* (Chicago: Museum of Science and Industry).

Lichtenberg, F.R., and D. Siegel, 1989. "The Effects of Leveraged Buyouts on Productivity and Related Aspects of Firm Behavior." Unpublished manuscript.

Lorell, M.A., 1980. *Multinational Development of Large Aircraft: The European Experience.* (Santa Monica, CA: RAND Corporation).

Malerba, F., 1985. *The Semiconductor Business.* (Madison, WI: University of Wisconsin Press).

Markoff, J., January 23, 1989. "A Corporate Lag in Research Funds is Causing Worry." *New York Times* A1.

Bibliography

Mattill, J., 1991. *The Flagship: The M.I.T. School of Chemical Engineering Practice, 1916–1991*. (Cambridge, MA: Koch School of Chemical Engineering Practice, M.I.T.).

McMillan, F.M., 1979. *The Chain Straighteners*. (London: Macmillan).

Millard, A., 1990. *Edison and the Business of Innovation*. (Baltimore, MD: Johns Hopkins University Press).

Miller, R., and D. Sawers, 1968. *The Technical Development of Modern Aviation*. (London: Routledge and Kegan Paul).

Morton, M., 1982. "History of Synthetic Rubber," in Raymond B. Seymour, ed., *History of Polymer Science and Technology*. (New York: Marcel Dekker).

Mowery, D.C., 1981. "The Emergence and Growth of Industrial Research in American Manufacturing, 1899–1946." (Ph.D. dissertation, Stanford University).

Mowery, D.C., 1983. "Industrial Research, Firm Size, Growth, and Survival, 1921–1946." *Journal of Economic History* 43, 953–980.

Mowery, D.C., 1984. "Firm Structure, Government Policy, and the Organization of Industrial Research: Great Britain and the United States, 1900–1950." *Business History Review* 58, 504–531.

Mowery, D.C., 1987. *Alliance Politics and Economics: Multinational Joint Ventures in Commercial Aircraft*. (Cambridge, MA: Ballinger Publishers).

Mowery, D.C., ed., 1988. *International Collaborative Ventures in U.S. Manufacturing*. (Cambridge, MA: Ballinger Publishing Company).

Mowery, D.C., 1995. "The Boundaries of the U.S. Firm in R&D," in N.R. Lamoureaux and D.M.G. Raff, eds., *Coordination and Information: Historical Perspectives on the Organization of Enterprise*. (Chicago: University of Chicago Press for the NBER).

Mowery, D.C., 1997. "The Bush Report After 50 Years: Blueprint or Relic?" in C.E. Barfield, ed., *Science for the 21st Century*. (Washington, DC: American Enterprise Institute).

Mowery, D.C., 1998. "The Computer Software Industry," in D.C. Mowery and R.R. Nelson, eds., *The Sources of Industrial Leadership*. (New York: Cambridge University Press).

Mowery, D.C., and N. Rosenberg, 1989a. *Technology and the Pursuit of Economic Growth*. (New York: Cambridge University Press).

Mowery, D.C., and N. Rosenberg, 1989b. "New Developments in U.S. Technology Policy: Implications for Competitiveness and International Trade Policy." *California Management Review* 27, 107–124.

Bibliography

Mowery, D.C., and N. Rosenberg, 1993. "The U.S. National System of Innovation," in R.R. Nelson, ed., *National Innovation Systems: A Comparative Analysis*. (New York: Oxford University Press).

Mowery, D.C., and W.E. Steinmueller, 1994. "Prospects for Entry by Developing Countries into the Global Integrated Circuit Industry: Lessons from the United States, Japan, and the NIEs, 1955–1990," in D.C. Mowery, ed., *Science and Technology Policy in Interdependent Economies*. (Boston: Kluwer Academic Publishers).

Mueller, W.F., 1962. "The Origins of the Basic Inventions Underlying Du Pont's Major Product and Process Innovations, 1920 to 1950," in *The Rate and Direction of Inventive Activity*. (Princeton, NJ: Princeton University Press).

National Research Council, 1982. "Research in Europe and the United States," in *Outlook for Science and Technology: The Next Five Years*. (San Francisco: W.H. Freeman).

National Resources Planning Board, 1942. *Research – A National Resource*, vol 1. (Washington, DC: U.S. Government Printing Office).

National Science Board, 1981. *Science Indicators, 1980*. (Washington, DC: U.S. Government Printing Office).

National Science Board, 1983. *Science Indicators, 1982*. (Washington, DC: U.S. Government Printing Office).

National Science Board, 1993. *Science and Engineering Indicators, 1993*. (Washington, DC: U.S. Government Printing Office).

National Science Board, 1996. *Science and Engineering Indicators, 1996*. (Washington, DC: U.S. Government Printing Office).

National Science Foundation, 1966. *Federal Funds for R, D, and Other Scientific Activities, Fiscal Years 1965, 1966, and 1967*. (Washington, DC: U.S. Government Printing Office).

National Science Foundation, Office of Computing Activities, 1990. "Director's Program Review," December 15, 1970. National Science Foundation, Washington, DC; duplicated.

National Science Foundation, 1985. *Science and Technology Data Book*. (Washington, DC: National Science Foundation).

National Science Foundation, 1987. *Research and Development in Industry, 1986*. (Washington, DC: National Science Foundation).

National Science Foundation, 1996. *National Patterns of R&D Resources: 1996*. (Washington, DC: National Science Foundation).

Navin, R., and M.V. Sears, 1955. "The Rise of a Market for Industrial Securities." *Business History Review* 29, 105–38.

Bibliography

Neal, A.D., and D.G. Goyder, 1980. *The Antitrust Laws of the U.S.A.*, 3d ed. (Cambridge, UK: Cambridge University Press).

Nelson, R.R., 1984. *High-Technology Policies: A Five Nation Comparison.* American Enterprise Institute.

Nelson, R.R., and G. Wright, 1994. "The Erosion of U.S. Technological Leadership as a Factor in Postwar Economic Convergence," in W.J. Baumol, R.R. Nelson, and E.N. Wolff, eds., *Convergence of Productivity.* (New York: Oxford University Press).

Nevins, A., and F.E. Hill, 1957. *Ford: Expansion and Challenge, 1915–1933.* (New York: Scribners).

Noble, D., 1977. *America by Design.* (New York: Knopf).

Norberg, A.L., and J.E. O'Neill, with contributions by K.J. Freedman, 1992. *A History of the Information Processing Techniques Office of the Defense Advanced Research Projects Agency.* (Minneapolis: Charles Babbage Institute).

Noyce, R., and M. Hoff, 1981. "A History of Microprocessor Development at Intel," *IEEE Micro* 1, 8–21.

Office of Management and Budget, Executive Office of the President, 1995. *Budget of the U.S. Government for Fiscal 1996.* (Washington, DC: U.S. Government Printing Office).

Office of Technology Assessment, U.S. Congress, 1981. *An Assessment of the United States Food and Agricultural Research System.* (Washington, DC: U.S. Government Printing Office).

Office of Technology Assessment, U.S. Congress, 1986. *Technology, Public Policy, and the Changing Structure of American Agriculture.* (Washington, DC: U.S. Government Printing Office).

Office of Technology Assessment, U.S. Congress, 1992. *Biotechnology in a Global Economy.* (Washington, DC: U.S. Government Printing Office).

Office of Technology Assessment, U.S. Congress, 1984. *Commercial Biotechnology: An International Analysis.* (Washington, DC: U.S. Government Printing Office).

Oi, W., 1988. "The Indirect Effect of Technology on Retail Trade," in R.M. Cyert and D.C. Mowery, eds., *The Impact of Technological Change on Employment and Economic Growth.* (Cambridge, MA: Ballinger).

Okimoto, D.I., 1986. "The Japanese Challenge in High Technology," in R. Landau and N. Rosenberg, eds., *The Positive Sum Strategy.* (Washington, DC: National Academy Press).

Okimoto, D.I., and G.R. Saxonhouse, 1987. "Technology and the Future of the Economy," in *The Political Economy of Japan*, vol. 1, K. Yamamura and Y. Yasuba, eds., *The Domestic Transformation*. (Stanford: Stanford University Press).

Organization for Economic Cooperation and Development (OECD), 1984. *Industry and University: New Forms of Co-operation and Communication*. (Paris: OECD).

Organization for Economic Cooperation and Development (OECD), 1989. *The Internationalisation of Software and Computer Services*. (Paris: OECD).

Orsenigo, L., 1988. *The Emergence of Biotechnology*. (London: Pinter Press).

Ostry, S., 1990. *The Political Economy of Policy Making: Trade and Innovation Policies in the Triad*. (New York: Council on Foreign Relations).

Owens, L., 1994. "The Counterproductive Management of Science in the Second World War," *Business History Review* 68, 515–576.

Parker, W.N., 1972. "Agriculture," in L.E. Davis *et al.*, eds., *American Economic Growth: An Economist's History of the United States*. (New York: Harper & Row).

Patel, P., and K. Pavitt, 1986. "Measuring Europe's Technological Performance: Results and Prospects," in H. Ergas, ed., *A European Future in High Technology?* (Brussels: Center for European Policy Studies).

Pavitt, K.L.R., 1990. "What We Know about the Management of Technology." *California Management Review* 32, 17–26.

Perry, W.J., 1986. "Cultivating Technological Innovation," in R. Landau and N. Rosenberg, eds., *The Positive Sum Strategy*. (Washington, DC: National Academy Press).

Perry, N.J., June 23, 1986. "The Surprising Power of Patents." *Fortune* 57–63.

Pharmaceutical Research and Manufacturers' Association, 1996. "1996 Survey: 284 Biotechnology Products in Testing." (Washington, DC: website www.phrma.org/charts/biochart.htm).

Pisano, G.P., W. Shan, and D.J. Teece, 1988. "Joint Ventures and Collaboration in the Biotechnology Industry," in D.C. Mowery, ed., *International Collaborative Ventures in U.S. Manufacturing*. (Cambridge, MA: Ballinger Publishing Company).

Pollack, A., April 10, 1990. "Technology Company Gets $4 Million U.S. Investment." *New York Times* C17.

Pursell, C., 1977. "Science Agencies in World War II: The OSRD and Its Challengers," in N. Reingold, ed., *The Sciences in the American Context*. (Washington, DC: Smithsonian Institution).

Bibliography

Raff, D.M.G., 1991. "Making Cars and Making Money: Economies of Scale and Scope and the Manufacturing behind the Marketing." *Business History Review 65*, 721–753.

Raff, D.M.G., and M. Trajtenberg, 1997. "Quality-Adjusted Prices for the American Automobile Industry: 1906–1940," in T.F. Bresnahan and R. J. Gordon, eds., *The Economics of New Goods*. (Chicago: University of Chicago).

Ravenscraft, D., and F.M. Scherer, 1987. *Mergers, Sell-Offs, and Economic Efficiency*. (Washington, DC: Brookings Institution).

Redmond, K.C., and T.M. Smith, 1980. *Project Whirlwind: History of a Pioneer Computer*. (Bedford, MA: Digital Press).

Rees, M., 1982. "The Computing Program of the Office of Naval Research, 1946–53." *Annals of the History of Computing 4*, 102–120.

Reich, L.S., 1985. *The Making of American Industrial Research*. (New York: Cambridge University Press).

Reich, R.B., and E. Mankin, 1986. "Joint Ventures with Japan Give Away Our Future." *Harvard Business Review 64*, 78–86.

Reid, P.P., 1989. "Private and Public Regimes: International Cartelization of the Electrical Equipment Industry in an Era of Hegemonic Change, 1919–1939," unpublished Ph.D. dissertation, Johns Hopkins School of Advanced International Studies.

Reid, T.R., 1984. *The Chip*. (New York: Simon and Schuster).

Reuters News Service, November 6, 1995. "Microsoft, Compaq Snare Computer Networking Firms."

Rodgers, T.J., January–February, 1990. "Landmark Messages from the Microcosm." *Harvard Business Review*, 24–30.

Rosenberg, N., ed., 1969. *The American System of Manufactures*. (Edinburgh: Edinburgh University Press).

Rosenberg, N., 1972. *Technology and American Economic Growth*. (New York: Harper).

Rosenberg, N., 1982. *Inside the Black Box: Technology and Economics*. (New York: Cambridge University Press).

Rosenberg, N., 1985. "The Commercial Exploitation of Science by American Industry," in K.B. Clark, R.H. Hayes, and C. Lorenz, eds., *The Uneasy Alliance*. (Boston: Harvard Business School).

Rosenberg, N., 1994. *Exploring the Black Box*. (Cambridge, UK: Cambridge University Press).

Rosenberg, N., 1996. "Uncertainty and Technological Change," in R. Landau, T. Taylor, and G. Wright, eds., *The Mosaic of Economic Growth*. (Stanford, CA: Stanford University Press).

Rosenberg, N., 1998a. "Technological Change in Chemicals: The Role of University-Industry Relations," in A. Arora, R. Landau, and N. Rosenberg, eds., *Chemicals and Long-Term Economic Growth*. (New York: John Wiley).

Rosenberg, N., 1998b. "The Role of Electricity in Industrial Development." *The Energy Journal* 19, 7–24.

Rosenberg, N., and R.R. Nelson, 1994. "American Universities and Technical Advance." *Research Policy* 24, 323–348.

Rosenberg, N., and W.E. Steinmueller, 1988. "Why Are Americans Such Poor Imitators?" *American Economic Review* 78, 229–234.

Rosenbloom, R.S., October 7, 1985. "The R&D Pioneers, Then and Now," presented at "The R&D Pioneers," Hagley Museum and Library, Wilmington, DE.

Rumelt, R.P., 1988. "Theory, Strategy, and Entrepreneurship," in D.J. Teece, ed., *The Competitive Challenge*. (Cambridge: Ballinger).

Sahota G.S., 1968. *Fertilizer in Economic Development: An Econometric Analysis*. (New York: Praeger).

Salter, M.S., and W.A. Weinhold, 1980. *Merger Trends and Prospects*, report for the Office of Policy, U.S. Department of Commerce, Washington, DC.

Sapolsky, H., 1990. *Science and the Navy*. (Princeton, NJ: Princeton University Press).

Saul, S.B., December 1962. "The Motor Industry in Britain to 1914." *Business History* 5, 22–44.

Schmookler, J., 1957. "Inventors Past and Present." *Review of Economics and Statistics* 39, 321–333.

Schmookler, J., 1962. "Changes in Industry and in the State of Knowledge as Determinants of Industrial Invention," in R.R. Nelson, ed., *The Rate and Direction of Inventive Activity*. (Princeton, NJ: Princeton University Press for the NBER).

Schumpeter, J.A., 1942. *Capitalism, Socialism and Democracy*. (New York: Harper & Row).

Schurr, S.H., C.C. Burwell, W.D. Devine, and S. Sonenblum, 1991. *Electricity in the American Economy*. (Westport, CT: Greenwood Press).

Sharp, M., 1989. "European Countries in Science-Based Competition: The Case of Biotechnology." Designated Research Center Discussion Paper #72, Science Policy Research Unit, University of Sussex, 1989.

Sheehan, J.C., 1982. *The Enchanted Ring*. (Cambridge, MA: MIT Press).

Shockley, W., 1950. *Electrons and Holes in Semiconductors*. (New York: Van Nostrand).

Bibliography

Shreve, R.N., and J.A. Brink, 1977. *Chemical Process Industries*. (New York: McGraw-Hill).

Slater, R., 1987. *Portraits in Silicon*. (Cambridge, MA: MIT Press).

Smith, J.K., and David Hounshell, August 2, 1986. "Wallace H. Carothers and Fundamental Research at Du Pont." *Science*, 436–442.

Smith, J.K., 1988. "World War II and the Transformation of the American Chemical Industry," in E. Mendelsohn, M.R. Smith, and P. Weingart, eds., *Science, Technology, and the Military*. (Boston, MA: Kluwer Academic Publishers).

Society of the Plastics Industry, various years. *Facts and Figures of the United States Plastics Industry*. (New York: Society of the Plastics Industry).

Solow, R.M., 1957. "Technical Change and the Aggregate Production Function." *Review of Economics and Statistics* 39, 312–320.

Spitz, P.H., 1988. *Petrochemicals: The Rise of an Industry*. (New York: John Wiley).

Steinmueller, W.E., 1996. "The U.S. Software Industry: An Analysis and Interpretive History," in D.C. Mowery, ed., *The International Computer Software Industry*. (New York: Oxford University Press).

Stern, G., August 13, 1997. "For Grease Monkeys, High Tech Is Changing Mechanics of Repair." *Wall Street Journal* A1.

Stigler, G.J., 1968. "Monopoly and Oligopoly by Merger," in G.J. Stigler, ed. *The Organization of Industry*. (Homewood, IL: Irwin).

Stokes, R., 1994. *Opting for Oil: The Political Economy of Technological Change in the West German Chemical Industry, 1945–61*. (New York: Cambridge University Press).

Sturchio, J.L., October 7, 1988. "Experimenting with Research: Kenneth Mees, Eastman Kodak, and the Challenges of Diversification," presented at "The R&D Pioneers," Hagley Museum and Library, Wilmington, DE.

Swann, J.P., 1988. *Academic Scientists and the Pharmaceutical Industry*. (Baltimore, MD: Johns Hopkins University Press).

Taylor, G.D., and P.E. Sudnik, 1984. *Du Pont and the International Chemical Industry*. (Boston: Twayne).

Thackray, A., 1982. "University-Industry Connections and Chemical Research: An Historical Perspective," in *University-Industry Research Relationships*. (Washington, DC: National Science Board).

Thackray, A., J.L. Sturchio, P.T. Carroll, and R. Bud, 1985. *Chemistry in America, 1876–1976: Historical Indicators*. (Dordrecht: Reidel).

Thorelli, H.B., 1954. *Federal Antitrust Policy*. (Baltimore, MD: Johns Hopkins University Press).

Bibliography

Tilton, J.E., 1971. *The International Diffusion of Technology: The Case of Transistors.* (Washington, DC: Brookings Institution).

Trescott, M.M., 1981. *The Rise of the American Electrochemicals Industry, 1880–1910.* (Westport, CT: Greenwood Publishers).

Tropp, H.S., ed., 1983. "A Perspective on SAGE: A Discussion." *Annals of the History of Computing* 5, 375–398.

U.S. Bureau of the Census, 1957. *Census of Manufactures, 1954.* (Washington, DC: U.S. Government Printing Office).

U.S. Bureau of the Census, 1987. *1987 Statistical Abstract of the United States.* (Washington, DC: U.S. Government Printing Office).

U.S. Bureau of the Census, 1995. *Statistical Abstract of the United States: 1995* (Washington, DC: U.S. Government Printing Office, 1995).

U.S. Bureau of the Census, 1997. *Statistical Abstract of the United States: 1997.* (Washington, DC: U.S. Government Printing Office).

U.S. Department of Agriculture, Crop Reporting Board, 1985. *Commercial Fertilizers.* (Washington, DC: U.S. Government Printing Office).

U.S. Department of Commerce, 1919. *Census of Manufactures: 1914*, volume II, *Reports for Selected Industries and Detailed Statistics for Industries, by States.* (Washington, DC: U.S. Government Printing Office), p. 753.

U.S. Department of Commerce, Bureau of the Census, *Statistical Abstract of the United States.* (Washington, DC: U.S. Government Printing Office).

U.S. Department of Commerce, 1960. *Historical Statistics of the United States.* (Washington, DC: U.S. Government Printing Office).

U.S. Department of Commerce, Bureau of Economic Analysis, December 1996. *Survey of Current Business*, vol. 76, Table C.1.

U.S. Department of Commerce, 1997. *U.S. Industry and Trade Outlook 1998.* (New York: McGraw-Hill).

U.S. Department of Commerce, Bureau of the Census, *Current Industrial Reports: Semiconductors, Printed Circuit Boards, an Other Electronic Components*, various years. (Washington, DC: U.S. Government Printing Office).

U.S. Department of Commerce, Bureau of the Census, 1975. *Historical Statistics of the United States: Colonial Times to 1970*, vol. 1. (Washington, DC: U.S. Government Printing Office).

U.S. Department of Health and Human Services, *NIH Data Book, 1993.* (Washington, DC: U.S. Government Printing Office).

U.S. Department of Transportation, 1985. *Highway Statistics Summary to 1985.* (Washington, DC: U.S. Government Printing Office).

Bibliography

U.S. International Trade Commission, 1995. *A Competitive Assessment of the U.S. Computer Software Industry.* (Washington, DC: U.S. International Trade Commission).

Utterback, J.M., and A.E. Murray, 1977. "The Influence of Defense Procurement and Sponsorship of Research and Development on the Development of the Civilian Electronics Industry." Center for Policy Alternatives working paper #77-5, M.I.T.

Vincenti, W., 1990. *What Engineers Know and How They Know It.* (Baltimore, MD: Johns Hopkins University Press).

Von Neumann, J., 1987. "First Draft of a Report on the EDVAC," 1945; reprinted in William Aspray and Arthur Burks, eds., *Papers of John von Neumann on Computing and Computer Theory.* (Cambridge, MA: MIT Press).

Weart, S., 1979. "The Physics Business in America, 1919–1940," in N. Reingold, ed., *The Sciences in the American Perspective.* (Washington DC: Smithsonian Institution).

White, L.J., 1971. *The Automobile Industry Since 1945.* (Cambridge, MA: Harvard University Press).

Whitehead, A.N., 1925. *Science and the Modern World.* (New York: Macmillan).

White House Science Council, 1988. *High-Temperature Superconductivity: Perseverance and Cooperation on the Road to Commercialization.* (Washington, DC: Office of Science and Technology Policy).

Wildes, K.L., and N.A. Lindgren, 1985. *A Century of Electrical Engineering and Computer Science at MIT, 1882–1982.* (Cambridge, MA: MIT Press).

Williams, T.I., 1982. *A Short History of 20th Century Technology.* (New York: Oxford University Press).

Williamson, O.E., 1975. *Markets and Hierarchies.* (New York: Free Press).

Williamson, O.E., 1985. *The Economic Institutions of Capitalism.* (New York: Free Press).

Wise, G., October 7, 1985. "R&D at General Electric, 1878–1985," presented at "The R&D Pioneers," Hagley Museum and Library, Wilmington, DE.

Wolf, J., May 31, 1989. "Europeans Fear Obstacles by U.S. on Advanced TV." *Wall Street Journal* A16.

Womack, J., D.T. Jones, and D. Roos, 1990. *The Machine That Changed the World.* (New York: Rawson Associates).

Wriston, W.B., 1992. *The Twilight of Sovereignty.* (New York: Charles Scribners Sons).

Index

201